MATHEMATICS
AN EXPLORATORY APPROACH

MATHEMATICS
AN EXPLORATORY APPROACH

Robert G. Stein
California State College, San Bernardino

McGRAW-HILL BOOK COMPANY
New York   St. Louis   San Francisco   Düsseldorf   Johannesburg   Kuala Lumpur
London   Mexico   Montreal   New Delhi   Panama   Paris   São Paulo   Singapore
Sydney   Tokyo   Toronto

MATHEMATICS: AN EXPLORATORY APPROACH

1234567890 SGDO 798765

This book was set in Times Roman by Black Dot, Inc. The editors were Stephen D. Dragin and Barry Benjamin; the designer was Jo Jones; the production supervisor was Dennis J. Conroy. The drawings were done by ANCO Technical Services.
The printer was Segerdahl Corporation; the binder, R. R. Donnelley & Sons Company.

Library of Congress Cataloging in Publication Data

Stein, Robert G.
    Mathematics: an exploratory approach.

    Bibliography: p.
    1. Mathematics—1961–   I. Title.
QA39.2.S72   510   74-20635
ISBN 0-07-060993-4

To my wife, Ronnie

CONTENTS

PREFACE

This book is neither an old-fashioned "cookbook" treatment nor its usual replacement, a "modern" course in mathematics emphasizing logical structure and deduction. Instead this book is intuitive and inductive, viewing mathematics primarily as a kind of activity rather than a body of knowledge. In spirit it is somewhere between the works of George Polya and W. W. Sawyer. Wherever possible it asks rather than tells, seeking to involve the student in raising mathematical questions and attempting to answer them. To this end many ideas are presented through experiments, both in the exercises and in the text itself. Naturally, this makes the book especially suitable for activity-oriented courses.

A particularly important consideration underlying this book is that many prospective elementary teachers dislike mathematics, are weak at it, and often try to avoid it. Therefore every effort has been made to present the material in a way that will involve the student without threatening him. In this the first lesson is crucial; it sets the tone for the entire book and strongly influences the frame of mind the student will bring to subsequent material. Thus instead of beginning with sets and numeration, as the logic of the material might suggest, this book starts with an open-ended exploration designed to involve the student and show him that he can indeed understand mathematics.

To maintain this attitude in later sections, even in more sophisticated material, jargon, symbolism, and abstraction are minimized, and generality and rigor are sacrificed for concreteness and clarity. Explanations rely freely on diagrams and are phrased in terms of particular cases instead of general cases. To reinforce the view of mathematics as a living subject, problems are mentioned whose solutions are not known today, and many historical references are included.

The approach to mathematics used in this book is consistent with recent work in the psychology of learning, which stresses the importance of introducing mathematical ideas concretely and progressing only gradually to abstractions. This means that the material is presented here in a way that is particularly useful for prospective teachers, because it can be readily adapted for use in elementary schools.

There is ample material here for a semester, but it can easily be adapted for courses lasting only a quarter. Perhaps the best way to do this is to treat the starred sections and problems as optional enrichment material, though students who will teach only the primary grades may skip Chaps. 12, 14, and 15 entirely, allowing a more leisurely treatment of the rest.

I owe a great debt of thanks to all those who helped me write this book. This includes many colleagues and students, especially Rochelle Campbell, who encouraged me, helped prepare preliminary versions for class testing, and suggested improvements. I must also thank the outstanding teachers I have known, especially Hans Hollstein and W. W. Sawyer, for their inspiration. Last and by far most, I thank my wife, Ronnie, who bore through the long project patiently and used her knowledge of teaching to suggest many improvements.

Robert G. Stein

TO THE READER

A famous college football coach once remarked, "Ninety percent of the job is getting the kids on your side." This applies to all teaching, but in mathematics it is easier said than done, because the subject is widely disliked. This book is designed to change that; one of its main goals is to show how elementary mathematics can be presented in an interesting way. This book is designed to build up your mathematical background; and if you teach, you can use it as a source of specific ideas for your own classroom.

Here you will find nothing to memorize and no repetitious drill. Instead you will find yourself challenged to explore mathematical questions, using your imagination and creativity. You may be surprised to find how much of mathematics you can discover for yourself. Of course exploration cannot be done passively: you are involved actively throughout. Read with pencil in hand, guessing at patterns and testing ideas as you go.

The exercises are a particularly important part of the book. Some are routine, designed to help you practice and test yourself on material discussed in the text. Others lead you to discoveries, amplifying the text or taking up points not mentioned there. Do no more of the routine exercises than you need for your own self-confidence. Concentrate instead on the more interesting exploration and pattern exercises. Do as many of these as you can, using any shortcuts and clever ideas that occur to you. (Ample practice in basic skills is built into these exploration exercises.) A star by a problem indicates that it is either especially challenging or a bit off the beaten path; skip it if it does not interest you. Similarly, starred portions of the text are included primarily as enrichment. They may be skipped with no loss of continuity, but you will probably find among them some of the most interesting and enjoyable parts of the book.

As you read and as you try to use ideas from this book in your own classroom, you will doubtless find places where it can be improved. Your suggestions and comments will be most welcome.

Robert G. Stein

CHAPTER 1
ADDING AND MULTIPLYING

Mathematics is an activity of exploration and investigation. This chapter will involve you in such activity. Do not be afraid if your arithmetic is rusty; the work in this chapter involves simple numbers and has built-in checks, and in the course of the book arithmetic is reviewed in some detail.

1.1 BOX PUZZLE EXPERIMENTS

Adding across the rows and down the columns of a little box like Fig. 1.1 leads to some interesting discoveries. Adding across the top row, we put 10, which is 2 + 8, in the upper right-hand corner and 12, which is 7 + 5, in the square just below the 10, as in Fig. 1.2. Next we add our original numbers down the columns to get Fig. 1.3, leaving only the lower right-hand corner empty. How should it be filled in?

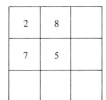

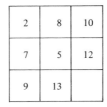
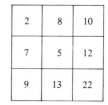

| Figure 1.1 | Figure 1.2 | Figure 1.3 | Figure 1.4 |

Adding down the right-hand column yields 10 + 12 = 22, and adding across the bottom row yields 9 + 13, which is also 22. Since both ways yield the same result, we complete the box as in Fig. 1.4.

That the box is now filled in may signify the end of the exercise, but the inquisitive always look for new questions to ask. Questions are the key to discovery. When adding down the right-hand column leads to the same number,

Figure 1.5

Figure 1.6

22, as adding across the bottom row, is this a numerical coincidence or something more fundamental? Since a good starting point for any investigation is to gather more evidence, we repeat the experiment with different numbers. Starting with Fig. 1.5 leads to Fig. 1.6. Now adding down the right-hand column yields 38, and so does adding across the bottom row.

There does indeed seem to be something to investigate. Is there any significance to the fact that both examples began with two even numbers and two odd numbers? What if one or more of the numbers were zeros? What if larger numbers or fractions are involved? (If you do not know how to add fractions, omit this part.) What if the format is changed, for example to something like the puzzles shown in Fig. 1.7? What if addition is replaced throughout by multiplication? Before reading on, try to answer these questions for yourself by making up examples and observing the results. Try to find out when the "coincidence" happens and when it does not, and why. When you have experimented and pondered a bit, read on.

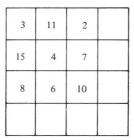

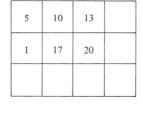

Figure 1.7

Are you convinced that more than coincidence is involved? Did you notice that when each box is complete, the number in the lower right-hand corner is the sum of the numbers with which the box was begun? (The exception occurs when addition is replaced by multiplication, and then the number is the result of multiplying the original numbers, which is called their *product*.) In our first example the number 22 is the sum of the original four numbers 2, 8, 7, and 5, though this sum was found two ways. Adding first across the rows and then down the right-hand column yields $(2 + 8) + (7 + 5)$. The parentheses indicate grouping; whatever is inside a pair of parentheses is to be dealt with as a single number, so that $(2 + 8) + (7 + 5)$ is really $10 + 12$. Adding first down

Figure 1.8

Figure 1.9

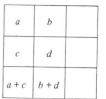

Figure 1.10

the columns and then across the bottom row yields $(2 + 7) + (8 + 5)$ or $9 + 13$.

Similar reasoning applies to other examples. If the four numbers are called a, b, c, and d, as in Fig. 1.8, adding across the rows yields Fig. 1.9, and adding down the columns yields Fig. 1.10. Adding down the right-hand column of Fig. 1.9 yields $(a + b) + (c + d)$, and adding across the bottom row of Fig. 1.10 yields $(a + c) + (b + d)$. Can you complete the reasoning from here? The key to the investigation is that $(a + b) + (c + d)$ is the same as $(a + c) + (b + d)$ no matter what numbers a, b, c, and d stand for. This is true because if only addition is involved, neither grouping nor order has any bearing on the result. The "coincidence" observed earlier will occur regardless of what numbers you start with because it depends on two basic properties of addition.

The fact that grouping does not affect a sum is known as the *associative* property of addition, since it deals with how numbers to be added are grouped, or associated. Symbolically, for any three numbers p, q, and r,

$$(p + q) + r = p + (q + r)$$

Because addition is associative, sums of more than two numbers may be written without grouping symbols since no matter how they are grouped the sum will be the same.

In our statement of the associative property, the numbers p, q. and r are written in the same order (left to right; first p, then q, then r) both times. That a sum is independent of the order in which the numbers are added is summarized in a second basic property, the *commutative* property of addition. Symbolically, if s and t are any numbers whatever, then

$$s + t = t + s$$

Addition is not the only process which is both associative and commutative. Multiplication is too. Therefore it is not surprising that the box puzzles still work when addition is replaced throughout by multiplication. The associative and commutative properties, like most "new mathematics," are old. Chrystal's "Textbook of Algebra," first published in 1882, was one of the first books to treat numbers in terms of these and a few other basic laws. It is a reasonable teaching approach, as it simplifies the subject from a collection of seemingly unrelated facts and operations to be memorized to a few general principles.

Recently this approach has been tried in elementary schools, but there have been problems in carrying it out. One is the confusion of words with the ideas for which they are supposed to stand. Most elementary school children find words like "commutative" and "associative" hard; they do not need such words. They do need to know, operationally, the concepts behind the words. That is, they need to know that order and grouping do not affect sums and products, and they need to know how to use these ideas in computation. Words should be introduced only where they help, and they should be kept simple and clear. That is the general approach in this book, though some words are introduced not because they are necessary in themselves but because they are widely used in today's elementary school books.

A direct, intuitive explanation of box puzzles for addition is available, as long as fractions, negative numbers, or other complications are avoided, for then addition can be interpreted in terms of merging collections of objects. (This operation, known technically as the *union of sets*, is discussed more fully in the Appendix.) Think of the numbers as representing numbers of objects in each square, so that Fig. 1.11 represents the puzzle in Fig. 1.1. With this interpretation, adding across the top row may be thought of as putting all the objects from the top row (the two circles and the eight crosses) into the upper right-hand corner. Similarly adding across the middle row will put the seven squares and five triangles into the right column, as in Fig. 1.12. Then adding down the right-hand column puts all the original objects into the lower right-hand corner, explaining why the number which appears there is the sum of the four numbers we started with. If instead we had begun by adding down the columns and then across the bottom row, we still would have ended up with all the original objects in the lower right-hand corner, showing why this number can be obtained either by adding across the rows first or down the columns first. This reasoning holds regardless of how many objects appear in each square. It even holds for box puzzles like those in Fig. 1.7, in which more squares are involved.

Box puzzles are an excellent way to teach the associative and commutative laws intuitively. At the same time they offer practice in computation, not as drill but as part of a larger investigation. Students need computation practice and enjoy it when it is not boring; in this case the children make up puzzles and experiments suitable to their own level of difficulty. The teacher should not spoil the fun by leading them to an explanation too soon, because it robs them of practice and prevents them from reaching their own understanding. Besides,

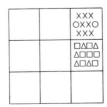

Figure 1.11   **Figure 1.12**

it takes students time to recognize that there is something worth investigating. Explaining too soon is like writing a whodunit and showing in chapter 1 that the butler did it. After that there is no point to the rest. We shall see some other "mysteries" about box puzzles, but these involve the basic relationship between multiplication and addition, which we pause to consider first.

Exercises 1.1

1 The commutative and associative properties of addition are often used (usually by people who do not realize it) to simplify mental arithmetic. For example, to add $87 + 71 + 13$, one might first add 87 to 71 to get 158, then add 13 to that to get 171. This amounts to grouping $87 + 71 + 13$ as $(87 + 71) + 13$. But if you notice that $87 + 13 = 100$, you can see right away that the sum is 171. The procedure can be justified by these successive applications of the associative and commutative properties:

Step 1: $(87 + 71) + 13 \quad = 87 + (71 + 13)$

Step 2: $\qquad\qquad\qquad = 87 + (13 + 71)$

Step 3: $\qquad\qquad\qquad = (87 + 13) + 71$

Which of these steps involve the associative property and which the commutative property?

2 Does the English language have the associative property? Ella Wheeler Wilcox, a journalist who took herself seriously as a poet, was furious when her line

My soul is a lighthouse keeper

appeared in print as

My soul is a light housekeeper

A more mundane example can be worked with "Is your doghouse broken?" Can you think of others?

1.2 ON MULTIPLICATION

In the simplest cases, which do not involve fractions or other complications, multiplication may be regarded as an abbreviation for repeated addition of the same number. For example, $3 + 3 + 3 + 3 + 3$ is 3 added 5 times, which is called "5 times 3" or "five 3s." Finding this sum is called "multiplying 5 by 3." A number which results from multiplication is called a *product*, and the numbers that are multiplied are called *factors*. In our example, addition shows that the product of 5 and 3 is 15.

In elementary schools the symbol $\times$ is usually used for multiplication, but in more advanced work a centered dot is often used, perhaps to avoid confusing $\times$ with x. We shall usually use the dot. Note that it is written high enough not to look like decimal point.

Counting the objects in a rectangular array is sometimes a helpful way to

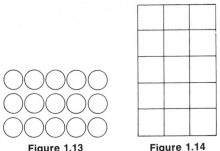

Figure 1.13   **Figure 1.14**

visualize a multiplication. For example, Fig. 1.13 shows a rectangular array with three rows of five dots each and so shows $5 + 5 + 5$ or $5 \cdot 3$. Counting the dots shows that $5 \cdot 3 = 15$. When the picture is turned on its side, it can be seen as five rows of three dots each, showing $3 + 3 + 3 + 3 + 3$ or $3 \cdot 5$. Since turning an array on its side does not change the number of dots in it, we can see that if p and q are whole numbers greater than 0, $p \cdot q = q \cdot p$, because an array of p rows with q dots in each can be turned on its side to form an array with q rows each with p dots. When we extend the concept of multiplication to other numbers, such as fractions and negative numbers, we shall see that no matter what numbers are multiplied, the order of the factors has no bearing on the product or, more technically, that multiplication is commutative for all numbers. If no sign is put between numbers, multiplication is conventionally understood, so that $p \cdot q$ may be shortened to pq or $2 \cdot a$ written $2a$. This shortcut is never used when it would lead to confusion. For example, if no sign were put between factors of 5 and 3, the product would look like the single number 53.

It will often be useful to interpret products geometrically in terms of area. Area is usually measured in terms of unit squares of a certain size, for example, a meter or a foot on each edge. If the length and width of a rectangle are both whole numbers of units, the region inside the rectangle can be cut up into a rectangular array of unit squares. The area of this region is then the product of the rectangle's length and width. Figure 1.14 shows a rectangle cut up this way, illustrating $5 \cdot 3 = 15$ in terms of area. This interpretation will clarify certain cases, such as multiplication of fractions, where arrays of dots do not help.

Exercises 1.2

1   What multiplications correspond to these pictures?

2 Draw arrays to picture:

(*a*) 3 · 7   (*b*) 2 · 1   (*c*) 5 · 8

*1.3 **AN INTERESTING APPLICATION**[1]

When Gauss was in elementary school it is said that his teacher told him to add up all the whole numbers from 1 through 100,[2] thinking it would keep the precocious child busy for a while. Instead Gauss did it correctly in a few minutes. The story is believable, because a shortcut to this and related problems makes them so simple that anyone who understands box puzzles can solve them. Children enjoy this trick because the method is ingenious and enables them to deal with problems which appear overwhelming at first. We begin by showing how the method can be used for the simple problem of adding the whole numbers from 1 through 10.

First we make a box puzzle, in which the numbers to be added are written in one order in the top row and in the reverse order in the second row, as in Fig. 1.15. When the puzzle is completed, each of the starred boxes will have the number we seek, which is $1 + 2 + 3 + \ldots + 9 + 10$ (the three dots indicate that the pattern is to be continued). Of course we could just add across to find this sum, which amounts to doing the problem the direct but naïve way. What if we first add down the columns and then across the bottom? This sounds difficult, but it turns out to be easy because the sum is the same in each column. (What is it?) Since there are 10 such columns, we can find what goes in the lower right-hand corner by multiplying. (What do you get?) Now the sum we seek, $1 + 2 + 3 + \ldots + 10$, belongs in each of the starred boxes, and so the number in the lower right-hand corner is twice that sum. From this we can find $1 + 2 + 3 + \ldots + 9 + 10$. (What is it?)

The key to the method is to relate the problem of adding numbers no two

[1]Starred sections may be skipped: see To the Reader, p. xiii.
[2]Karl Friedrich Gauss (1777–1855) was one of the greatest mathematicians of all time. He was born in Brunswick, Germany, where his talents soon attracted the attention of a duke who financed his education. At the age of seventeen his discovery on the constructibility of polygons led him to give up his idea of becoming a philologist and become a mathematician instead. In 1807 he was named director of the Göttingen observatory, and he held this post until he died. He made immense contributions to all parts of mathematics as well as to astronomy and to the study of magnetism.

1	2	3	4	5	6	7	8	9	10	*
10	9	8	7	6	5	4	3	2	1	*

Figure 1.15

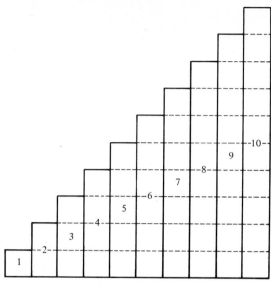

Figure 1.16

of which are equal to the simpler problem of repeated addition of the same number, which is multiplication. The method thus depends on the fact that when the numbers to be added are written forward and backward, the corresponding numbers add to a constant sum (11 in the above example). Geometrically, if a number like 3 is interpreted as the area of a rectangle 1 by 3, then the sum to be found is the area of the step shape in Fig. 1.16. The indirect way of computing this area amounts to putting a duplicate set of steps upside down on the first one to form a rectangle (as in Fig. 1.17). The area of the rectangle can be found by multiplication, and then taking half that area gives the area of the original step figure.

This method can be used as a shortcut to add any sequence of numbers with the property that when it is written forward and backward the corresponding numbers have a constant sum. Gauss' problem of summing the numbers from 1 to 100 is a simple example. Imagine a long box puzzle with the whole numbers from 1 through 100 written across the top row in ascending order and in the opposite order across the second row. Such a puzzle would be awkward to write, but imagine it to have been put on a long scroll of paper, which in Fig. 1.18 is shown folded to fit on the page. Once again the number we seek will eventually go in each of the starred boxes, but in this case it would be a very tedious job to add across either of these rows. Adding down the columns shows them to have the constant sum 101. Because the numbers being added go up in constant steps, we can be sure that even the columns which cannot be seen in Fig. 1.18 add to 101. There are 100 such columns, so that the number which belongs in the lower right-hand corner of this long puzzle is $101 \cdot 100$ or 10,100. The starred numbers are each half of this, or 5,050, which is therefore $1 + 2 + 3 + \ldots + 99 + 100$.

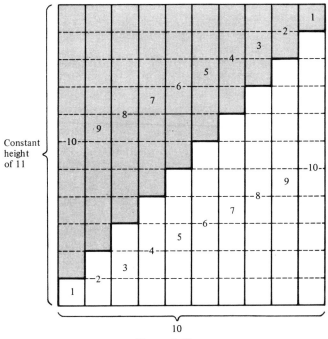

Figure 1.17

The beauty of this method is that with it the sum of the first million numbers is really no harder to find than the sum of the first 100. Of course a box puzzle a million squares long staggers the imagination, but its ends would appear as in Fig. 1.19 (the middle has been torn out). Once again the columns add to a constant, which in this case is 1,000,001. There are a million such columns, and so the box in the lower right-hand corner will contain

$$1,000,001 \cdot 1,000,000 = 1,000,001,000,000$$

Each of the starred boxes in this case contains half of this, or 500,000,500,000,

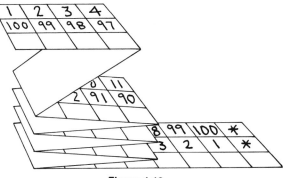

Figure 1.18

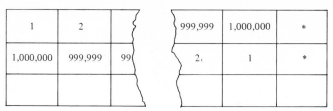

Figure 1.19

which is therefore the sum of the whole numbers from 1 through 1 million.

This method shows that a little thought can save a lot of work, but it also shows a way to deal with numbers so big that they would be impossible to count. We could use it to add the numbers from 1 through 1 trillion in a few minutes, even though 1 trillion is such a huge number that the United States has not yet existed for that many seconds!

Our indirect method of addition can be used to add any sequence of numbers in which the increase or decrease from each to the next is constant. These are called *arithmetic sequences* or *arithmetic progressions*. Some examples of arithmetic sequences are

(a)   51, 53, 55, 57, 59, . . . , 97, 99
(b)   3, 6, 9, 12, 15
(c)   110, 120, 130, 140, 150, 160
(d) $\frac{1}{2}$, 1, $1\frac{1}{2}$, 2, $2\frac{1}{2}$, 3, $3\frac{1}{2}$, 4, . . . , $9\frac{1}{2}$, 10

For example, to add the numbers in sequence (a), the odd numbers between 50 and 100, the box puzzle looks like Fig. 1.20, again with the middle torn out. Here the columns add to the constant sum 150. How many columns are there? This subsidiary problem involves some careful reasoning.

One approach is to observe that there is one column for each odd number between 50 and 100. Of the whole numbers from 1 through 100, the first 50 are not involved. Half of the second 50 are odd, and so there are 25 numbers in the sum $51 + 53 + \cdots + 97 + 99$ and 25 columns in the puzzle of Fig. 1.20. The doubly starred box in Fig. 1.20 should therefore be filled in with $25 \cdot 150 = 3{,}750$, and so in each singly starred box there belongs half of this, or 1,875. Thus $51 + 53 + \cdots + 99 = 1{,}875$.

51	53	55	57		95	97	99	*
99	97	95	9		5	53	51	*
								**

Figure 1.20

Exercises 1.3

1   Verify by direct addition that $1 + 2 + 3 + \cdots + 9 + 10 = 55$.

2   Add:

(a) $1 + 2 + 3 + \cdots + 199 + 200$

(b) $4 + 6 + 8 + 10 + \cdots + 52 + 54 + 56$

(c) $151 + 150 + 149 + \cdots + 51 + 50 + 49$ (Does it matter that the numbers are written in descending order?)

(d) $\frac{1}{2} + 1 + 1\frac{1}{2} + 2 + 2\frac{1}{2} + 3 + \cdots + 8\frac{1}{2} + 9 + 9\frac{1}{2}$. If you do not know how to add fractions, skip this.

(e) $103 + 106 + 109 + \cdots + 151 + 154$

(f) $1 + 2 + 3 + \cdots + 999{,}999{,}999 + 1{,}000{,}000{,}000$

3   A man walked 100 yards on January 1, 200 yards on January 2, 300 yards on January 3, and kept increasing his stroll by 100 yards each day for the rest of the year, which was not a leap year. What was the total length of these walks for the year?

4   Add the whole numbers between 1 and 100 which end in either 2 or 8; that is, add $2 + 8 + 12 + 18 + 22 + \cdots + 88 + 92 + 98$.

  Note: The increases between successive numbers being added are alternately 6 and 4. The discussion above was confined to the case where the increases are constant, but the ideas can be used here. The following hints give two ways to do this.

(a) Separate the sum to be found into two parts, one the sum of those numbers which end in 2 and the other the sum of those numbers which end in 8. (What do you observe about $2 + 12 + 22 + \cdots + 82 + 92$ and $8 + 18 + 28 + \cdots + 88 + 98$?)

(b) Disregard the fact that the increases are not all the same and just write the numbers forward and backward, in a box puzzle. What do you observe? Can you explain?

*5   Add the whole numbers between 0 and 1,000 which end in 1, 3, 7, or 9. (Remember, the star means the problem is challenging; in this case there are several good approaches.)

1.4 DISTRIBUTIVITY

Figure 1.21 shows an interesting variation on box puzzles. The four numbers at the upper and left-hand sides are keys. To fill in each starred square, multiply the border numbers for its row and column, as in Fig. 1.22. There, for example, the center box contains 14, which is the product of the 7 and the 2 on the border. The remaining squares are filled by addition, as in the earlier puzzles, and then the remaining border numbers are found by adding along the borders, as shown by the arrows in Fig. 1.22. The completed puzzle is shown in Fig. 1.23. Look at it for a while.

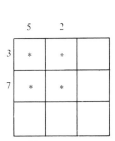

Figure 1.21

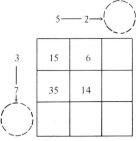

Figure 1.22

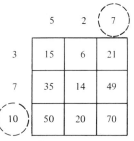

Figure 1.23

The addition part of the box is no longer a surprise, but you should find something else of interest. In particular note the relationship of the number in the lower right-hand square to the two circled border numbers. What questions should you ask next? Try to frame good questions and answer them by experiment, before reading on. (If you have trouble thinking of good questions, look at those in Sec. 1.1. Your experiments should include making up several such puzzles and trying to find when and why they work. It may help you to visualize products in terms of area.)

These puzzles are easy to understand by interpreting the products as area. Figure 1.24 shows a rectangular area partitioned into four smaller rectangles whose lengths and widths in centimeters are the border numbers of the puzzle

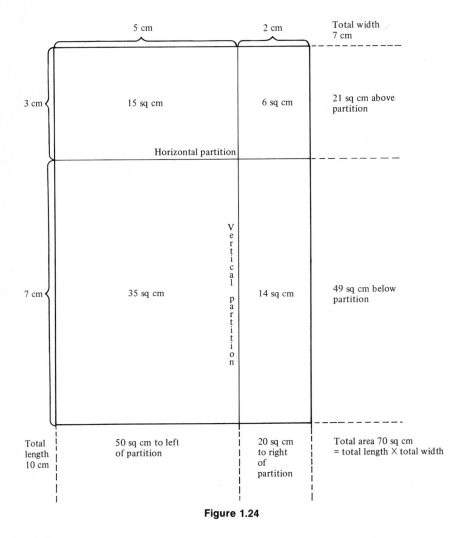

Figure 1.24

in Fig. 1.21. Suppose we want to find the total area of this rectangle. We could first find the areas of the four smaller rectangles and then add them, or we could calculate the total length and total width of the entire rectangle and multiply them. The resulting numbers in Fig. 1.24 are just the same as those in Fig. 1.23. A similar scale drawing could be made for any box puzzle with border numbers.

Underlying this is the *distributive* property, which relates addition and multiplication by the following basic rule. For any three numbers a, b, and c,

$$a(b + c) = ab + ac$$

(*Reminder*: Where no operation is indicated, multiplication is understood.) Sometimes this rule is expressed $(b + c)a = ba + ca$, but since order does not affect products, this is really no different.

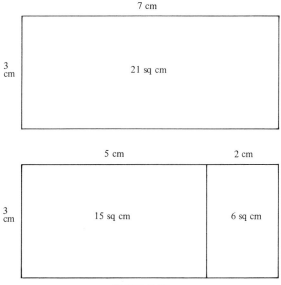

Figure 1.25

To see how the distributive property is involved in the box puzzles, consider Fig. 1.25, which is that part of Fig. 1.24 above the horizontal partition. The area above the partition is unchanged whether or not one uses the partition in calculating it; this is shown by $21 = 15 + 6$. When each number is expressed as a length times a width, this is $3 \cdot 7 = 3 \cdot 5 + 3 \cdot 2$. Since $7 = 5 + 2$, this amounts to $3(5 + 2) = 3 \cdot 5 + 3 \cdot 2$, which is the distributive property with $a = 3$, $b = 5$, and $c = 2$.

A word about parentheses. One might interpret an expression like $4 \cdot 8 + 9$ either as $32 + 9 = 41$ or as $4 \cdot 17 = 68$, depending on whether the addition or the multiplication is done first. Either interpretation is reasonable, but the ambiguity is not. To avoid it mathematicians everywhere have adopted the convention that in expressions which involve both addition and multiplication, as this one does, the multiplications are to be carried out before the additions unless otherwise indicated. By this convention $4 \cdot 8 + 9$ means $32 + 9$, or 41. To indicate that the factor 4 also applies to the 9, you would write $4(8 + 9)$.

The above puzzle is merely a double application of the distributive property. On the one hand, the product $(3 + 7)(5 + 2)$ is $10 \cdot 7$, or 70. On the other hand, it is $3(5 + 2) + 7(5 + 2)$ by the distributive property. These can be multiplied out as $21 + 49$, or they can be further expanded by the distributive property to $3 \cdot 5 + 3 \cdot 2 + 7 \cdot 5 + 7 \cdot 2$ or $15 + 6 + 35 + 14$, numbers which are familiar by now.

Exercises 1.4

1 The top row of the puzzle in Fig. 1.23, $3 \cdot 5 + 3 \cdot 2 = 3(5 + 2)$, illustrates the distributive property. Find five more such illustrations in rows and columns of the same puzzle.

2 The distributive property can be extended to multiplications involving sums of three and more numbers. In that case it might be written $a(b + c + d) = ab + ac + ad$. See how this is involved with the following puzzles:

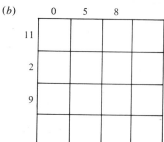

(*a*)   4   7   9   (*b*)   0   5   8

8   11

3   2

2   9

1.5 WHY WE MULTIPLY THE WAY WE DO

If a multiplication problem is too long to do in our head, we usually do it digit by digit. The process is quite simple, but many people know it only by rote. As an example, consider the multiplication of 351 by 82. The computation, shown in Fig. 1.26, is usually accompanied by something like this: "2 times 1 is 2, so we put it down; 2 times 5 is 10, so we put the 0 to the left of the 2 and remember the 1; 2 times 3 is 6, which, with the 1 we remembered, makes 7, and this goes to the left of the 0; now 8 times 1 is 8, which is put below the 0 of 702; 8 times 5 is 40, so we put down the 0 to the left of the 8 and remember the 4; 8 times 3 is 24, which, with the 4 we remembered is 28; adding, we find that $351 \cdot 82 = 28{,}782$."

This process is just a thinly disguised application of the distributive property. To see how it works, write 351 as $300 + 50 + 1$ and 82 as $80 + 2$ and make the box puzzle in Fig. 1.27. Adding the border numbers shows that the circles will contain 82 and 351. Their product will appear in the lower right-

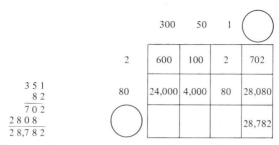

$$
\begin{array}{r}
3\ 5\ 1 \\
8\ 2 \\
\hline
7\ 0\ 2 \\
2\ 8\ 0\ 8 \\
\hline
2\ 8{,}7\ 8\ 2
\end{array}
$$

Figure 1.26   **Figure 1.27**

hand corner. Filling in the boxes and adding horizontally, we find that

$$2 \cdot 351 = 702 \quad \text{and} \quad 80 \cdot 351 = 28{,}080.$$
Thus
$$82 \cdot 351 = (80 + 2)351 = 80 \cdot 351 + 2 \cdot 351 = 28{,}080 + 702 = 28{,}782$$

The 702 and 28,080 are the same numbers that were added in Fig. 1.26, though the final 0 in 28,080 was omitted there. There is no reason for omitting the 0 except custom. Since adding 0 does not affect a sum, it does not matter whether it is included or not. Presumably children were taught to leave it out to save writing and time. It is questionable whether this ever saved as much time as it took to teach it. Every detail of the multiplication process quoted above can be found in Fig. 1.26. For example, the statement "2 times 5 is 10" refers to the multiplication of the 5 in 351 by 2. In fact the 5 in 351, because of its position, represents 50, and so this really states that $2 \cdot 50 = 100$, a fact which appears clearly in Fig. 1.27. Long multiplication of 351 by 82 amounts to the following:

$$
\begin{aligned}
82 \cdot 351 &= (80 + 2)351 = 80 \cdot 351 + 2 \cdot 351 \\
&= 80(300 + 50 + 1) + 2(300 + 50 + 1) \\
&= 80 \cdot 300 + 80 \cdot 50 + 80 \cdot 1 + 2 \cdot 300 + 2 \cdot 50 + 2 \cdot 1 \\
&= 24{,}000 + 4{,}000 + 80 + 600 + 100 + 2 \\
&= 28{,}782
\end{aligned}
$$

which is simply repeated application of the distributive law. The same is true of any long multiplication.

Exercises 1.5

1 In Fig. 1.27 the first three boxes in the bottom row are not filled in. Fill them in, and find out how those three numbers relate to computing the product of 351 and 82. *Hint:* Carry out the multiplication

$$\begin{array}{r} 82 \\ \times\,351 \\ \hline \end{array}$$

This is the same multiplication as in the text, but the order of the factors is reversed.

2 What long multiplications do these box puzzles represent? Complete the puzzles and the multiplications, and compare the results.

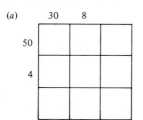   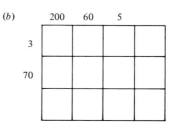

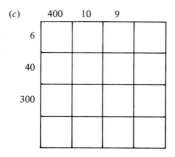

3 Make box puzzles to illustrate

   (a) 654 · 103   *(b) 781 · 1,729

   Then use the puzzles to find the products.

CHAPTER 2
INTRODUCTION TO NEGATIVE NUMBERS

Since people first began to count, the concept of *number* has been extended and refined. One such extension is to numbers "below zero."

Numbers below zero are known as *negative* numbers, a name with the unfortunate connotation of a value judgment. (A negative attitude, for example, is generally considered bad.) The medieval church associated negative numbers with debt and therefore regarded them as sinful. This effectively banned their use until the Renaissance, and even then negative numbers were written in the devil's color, red, a tradition which carried into modern bookkeeping.

2.1 THE NUMBER LINE

We so often think of numbers as points on a line that you probably take the association for granted when you use a ruler or a thermometer. The correspondence of numbers with points, simple though it is, has far-reaching consequences. Imagine an inchworm who inches himself along the number line in Fig. 2.1. Starting at 0, his first move takes him to 1, his second to 2 and so on. Since a line is assumed to be endless, as he continues, he determines a point on it for each number in the endless list 0, 1, 2, 3, 4, 5, ...

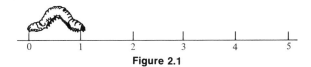

Figure 2.1

What if the inchworm moves in the other direction from 0? Now his first move lands not at the point which we have labeled 1 but at a new point, which we label ⁻1 (read "negative one"). Continuing, he comes to points ⁻2, ⁻3, ⁻4, ⁻5, . . . , as shown in Fig. 2.2. These numbers on the opposite side of 0 from 1

Figure 2.2

are called *negative*. By contrast, numbers on the same side of 0 as 1 are called *positive*; 0 is neither positive nor negative.

In Fig. 2.2, 0 is midway between 1 and ⁻1. It is also midway between 2 and ⁻2, 3 and ⁻3, and in fact between any number and its negative. In general, if *n* is a number, ⁻*n* is the number symmetrical to *n* on the other side of 0; we say ⁻*n* is the *opposite* of *n*. The raised minus sign thus means "take the opposite." This also applies to an expression like ⁻(⁻3) which is the opposite of the opposite of 3, namely 3 itself.

For now we shall use only whole numbers, called *integers*. In later chapters we shall extend out consideration to other numbers, such as fractions.

The two most popular sources of elementary examples involving negative numbers are probably subzero temperatures and deficit finance, because it is not easy to find concrete illustrations of negative numbers in settings familiar to elementary school children. Do not worry about this. Numbers are abstract tools of thought, but one can hardly expect the same tool to serve in all cases. For the teacher this means that negative numbers are best introduced abstractly. Fortunately, the popularity of bridge, chess, crossword puzzles, etc., shows that practicality is not the only motive for abstract thought, and classroom experience bears this out. Later we shall find many uses for negative numbers.

Exercises 2.1

1 Each of these phrases involves negative numbers somewhat artificially. Restate each in more usual language remembering to think of negative as "opposite."
(*a*) Our profit last month was negative.
(*b*) The meeting starts in ⁻5 minutes.
(*c*) The airplane is climbing at ⁻200 feet per minute.
(*d*) In a recession the economy may have a negative rate of growth.
(*e*) The rocket is approaching earth at ⁻3 miles per second.
(*f*) The north poles of two magnets attract each other negatively.
(*g*) A negative income tax for poor people.
(*h*) The quarterback gained ⁻7 yards.
2 In the text it is mentioned that ⁻(⁻3) = 3, since ⁻(⁻3) is the opposite of ⁻3.
(*a*) By the same reasoning, what is ⁻[⁻(⁻3)] in simpler terms?
(*b*) What is ⁻{⁻[⁻(⁻3)]}?
(*c*) What is ⁻[⁻(⁻{⁻[⁻(⁻3)]})]?
(*d*) Find a rule to explain how to tell from the number of minus signs whether expressions like those in parts (*a*), (*b*), and (*c*) are positive or negative.

2.2 HOW ARE NEGATIVE NUMBERS ADDED?

The most elementary concept of addition comes from putting collections of objects together. If there are n objects in one pile and m in another, putting the two piles together yields a new pile with $n + m$ objects. Unfortunately, however, this sheds no light on adding negative numbers, since no pile can contain a negative number of objects.

Figure 2.3

Figure 2.3 suggests another approach. Can you use it to read $2 + 5$? How about $2 + 4$? What other sums can you read? Extending the number lines, can you modify Fig. 2.3 to read $2 + 10$? Can you extend the number lines in Fig. 2.3 to make it show $2 + {}^-1$? How about $2 + {}^-2$ and $2 + {}^-3$? So far our sums have 2 as their first addend, a fact reflected in Fig. 2.3 by offsetting the number lines so that the top 0 is by the bottom 2. Now sketch a figure like Fig. 2.3 which shows sums with 3 as the first addend. You should be able to read from it $3 + 5$, $3 + 4$, etc., as well as $3 + {}^-1$, $3 + {}^-2$, $3 + {}^-3$, and $3 + {}^-4$.

This suggests the possibility of a slide-rule sort of device for adding. To make one, take a piece of squared paper (four or five lines to the inch is convenient) and fold it parallel to one edge, as in Fig. 2.4. Then cut off the top as indicated and place it upside down in the fold. Number the two pieces (using the lines for accurate spacing), as in Fig. 2.5.

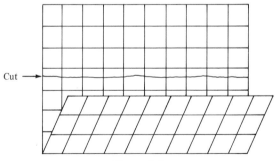

Figure 2.4

Now you can slide the inner piece to produce a variety of configurations, among them the one in Fig. 2.5. Experiment with this slide rule. Compute several sums, including some with both numbers positive, some with one number positive and the other negative, and some with both numbers negative.

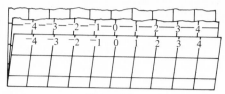

Figure 2.5

If you are unsure of yourself, verify these:

$$^-2 + ^-3 = ^-5 \qquad ^-1 + 5 = 4 \qquad ^-4 + 3 = ^-1$$

2.3 LEAVING THE NUMBER LINE

In practice you must be able to add positive and negative numbers without relying on a mechanical device. Here is one such method. It requires a little imagination, but with practice you will find it easy.

Imagine two rival armies, the "positives" and the "negatives." Whenever a group of positives meets a group of negatives, they kill each other off one for one until one group or the other is annihilated. Thus if 3 negatives meet 7 positives, the negatives will kill off 3 of the positives, leaving 4 positives. This illustrates $^-3 + 7 = 4$. Again, if 50 positives encounter 75 negatives, there will be 25 negatives left after the dust has cleared, so that $50 + ^-75 = ^-25$. On the other hand, if a band of 7 negatives meets 9 more negatives, there will be no conflict. Instead they combine into a group of 16 negatives, which shows that $^-7 + ^-9 = ^-16$. Likewise if 10 positives meet 13 more positives, they reinforce each other to form a group of 23 positives in all, so that $10 + 13 = 23$. This may not sound "mathematical," but it accurately represents addition of positive and negative numbers.

Exercises 2.3

1 Add:
   (a) $^-5 + 5$ (b) $^-19 + 19$ (c) $27 + ^-27$ (d) generalize
2 Add (in your head if you can):
   (a) $^-12 + ^-27$ (b) $91 + ^-92$ (c) $^-48 + 1$ (d) $^-73 + ^-2$
   (e) $^-194 + 195$ (f) $^-72 + 71$ (g) $^-931 + 0$ (h) $^-80 + 80$
   (i) $3 + ^-61$ (j) $^-190 + 100$ (k) $^-1 + ^-100$ (l) $^-40 + ^-43$
3 Do box puzzles work with negative numbers? Try these, and make up more if you wish.

(a)

9	$^-4$	
$^-7$	3	

(b)

$^-21$	47	
$^-73$	52	

(c)

$^-9$	$^-5$	18	
16	$^-11$	29	
0	14	$^-31$	

4 In view of Prob. 3, do you think that the associative and commutative laws of addition hold when negative numbers are involved?

5 Suppose in Prob. 3 each of the given numbers in the box puzzles is replaced by its opposite. What effect will this have on the numbers to be filled in?

6 (a) Add the integers from ⁻10 through 10 (inclusive). Do this problem at least two ways, using one to check the other. *Hint*: Recall Sec. 1.3.

(b) Add all integers from ⁻25 through 35 inclusive.

(c) Add 26 + 27 + 28 + ⋯ + 35. (What do you notice? Why?)

(d) Add the odd integers between ⁻40 and 70. (Can you think of two ways to do this?)

7 Compute and compare:

(a)   ⁻(3 + 5) =   and   ⁻3 + ⁻5 =

(b)   ⁻(7 + 11) =   and   ⁻7 + ⁻11 =

(c)   ⁻(6 + ⁻4) =   and   ⁻6 + 4 =

(d)   ⁻(⁻2 + 4) =   and   2 + ⁻4 =

(e)   ⁻(⁻11 + ⁻5) =   and   11 + 5 =

(f) generalize

2.4 THE ADDER

Figure 2.6 shows a simple device for adding, consisting of three parallel rows of numbered dots. Can you figure out how to use this adder? Experiment for a while. What questions occur to you? When you have pondered a while, read on.

If you had trouble figuring out how to use the adder, try this hint: draw a line from the dot numbered 3 on one side to the dot numbered 5 on the other. Where does the line cross the center column? Try a few more like this.

Any discovery raises new questions, and the adder raises several. Why does it work? Would it work for larger numbers? For fractions? For negative numbers? Can it or something like it be made to subtract, multiply, or divide?

To see why the adder works, consider a concrete case, 3 + 2. Lining up the dots numbered 3 and 2 on opposite sides yields the shape in Fig. 2.7. The left-hand edge is 2 units long, and the right hand edge is 3 units long.

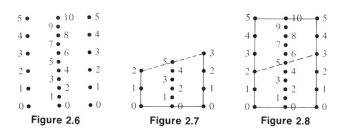

Figure 2.6   Figure 2.7   Figure 2.8

Now put a second shape just like it on top, as in Fig. 2.8. Together, the two shapes form a rectangle, whose height is, naturally, 3 + 2 units. The sloping line cuts the middle column of dots half way up, which will be $2\frac{1}{2}$ units from the bottom of the rectangle if we measure with the same units as on the outer scale. But the numbers in the center column are spaced only half as far apart as those in the outer column, so that the sloping line crosses the center column at the dot numbered 5 instead of $2\frac{1}{2}$.

The possibility of building a multiplier like our adder is fascinating, but since it leads to some rather advanced mathematics, it is postponed until Chap. 12. The exercises below deal with some other questions about the adder.

Exercises 2.4

1 Make diagrams like Fig. 2.7 and 2.8 to show how the adder correctly computes $5 + 2$.

*2 Even though we have not discussed fractions so far, you might want to experiment with a few simple fractions on the adder. For example, could you make it add $\frac{1}{2} + \frac{1}{2}$? How about $\frac{1}{2} + 1\frac{1}{2}$? Try adding other fractions.

3 Extend the dot patterns below 0 on the adder, number them, and experiment with adding positive and negative numbers on the adder. Do the results agree with earlier work? *Caution*: For the adder to work well, the dots must be spaced accurately. If you build your own, you will find squared paper a big help.

4 Two third graders set out to build a large adder by taping pieces of graph paper together. They hoped their adder would save them from ever adding by hand again. They built an adder that adds numbers up to 200, but then they gave up the project because their adder was not accurate. What are the main problems in making a large and accurate adder?

2.5 MULTIPLYING WITH NEGATIVE NUMBERS

Multiplication of positive integers is repeated addition. Can we stretch this view to multiplication with negative numbers? A few cases offer hope. For example, we may try $3(^-5) = ^-5 + ^-5 + ^-5$ (how much is that?) and $^-4 \cdot 2 = ^-4 + ^-4 = ^-8$. Is this reasonable? Do the patterns that hold for multiplication of positive integers carry over? The most important of these patterns, the distributive property, underlies the box puzzles with border numbers. Try these:

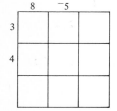 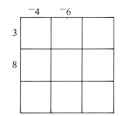 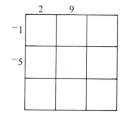

These should give you confidence that, provided at least one of the two factors is positive, multiplication can indeed be interpreted as repeated addition. (We shall deal with the product of two negative numbers shortly.)

Repeated addition is adequate when the numbers involved are simple, but what about cases like $^-79 \cdot 381$? It would be tedious to have to add $^-79$ to itself 381 times. Instead, examine the products you found above. How does $3(^-5)$ relate to $3 \cdot 5$? How does $4(^-5)$ relate to $4 \cdot 5$? How do $3(^-4)$, $3(^-6)$, and

3($^-$10) relate to 3 · 4, 3 · 6, and 3 · 10? The pattern is clear, and we shall adopt the following definition based on it:

For any numbers m and n, we define (^-m)n and m(^-n) to be $^-$(mn).

In words this means that negating a factor negates the product.

What if both factors are negative? For instance, what is ($^-$4)($^-$5)? Repeated addition makes no sense here (what would it mean?), but the box puzzles still work. Try this one:

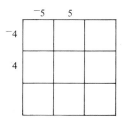

Save the upper left-hand corner for last so that it can be figured out indirectly. You will see that ($^-$4)($^-$5) is a number which can be added to $^-$20 to get 0. Do you know of such a number? (*Hint*: See Prob. 1 of Exercises 2.3.) In Chap. 6 we shall see that there can be only one such number. Make up a box puzzle to find ($^-$7)($^-$3).

People sometimes find it strange that the product of two negative numbers is positive, but they forget that the negative of a number is its opposite. A product of positive factors is positive. Negating one factor makes it negative, negating another makes it positive again, and so on. The situation is a bit like that in Prob. 2 of Exercises 2.1.

Exercises 2.5

1 Which of these products are positive? Which are negative?
   (*a*) ($^-$5) ($^-$13)   (*b*) ($^-$71) (40) ($^-$11)   (*c*) ($^-$4) ($^-$4) ($^-$4)
   (*d*) 7($^-$2) (8) ($^-$3) ($^-$9) ($^-$4) 10($^-$5)
   (*e*) 5($^-$4) (3) ($^-$2) (0) (1) ($^-$3)

2 (*a*)   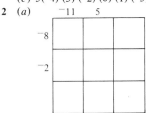   (*b*)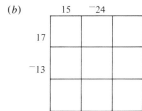

   If you want more practice, make up more of these box puzzles.
3 If $p = {}^-7$ and $q = 6$, how much are:
   (a) pq (b) $({}^-p)q$ (c) $p({}^-q)$ (d) $p + q$
   (e) $(p + q)q$ (f) $(p + q)p$ (g) $(p + q)(p + q)$
4 Many dreary hours have been spent by students trying to learn the multiplication table. Few
   people realize that it can be extended to include products with negative factors. To do this,
   write the factors in two intersecting lines, as shown in Fig. 2.9, and then fill in the products. For
   example, A is the product of ${}^-3$ and 4, while B is $({}^-5)({}^-4)$. Fill in this table, paying attention
   to the interweaving patterns.

								7							
								6							
								5							
				A				4							
								3							
								2							
${}^-7$	${}^-6$	${}^-5$	${}^-4$	${}^-3$	${}^-2$	${}^-1$	0	1	2	3	4	5	6	7
								0							
								${}^-1$							
								${}^-2$							
								${}^-3$							
		B						${}^-4$							
								${}^-5$							
								${}^-6$							
								${}^-7$							

Figure 2.9

CHAPTER 3
FIGURATE NUMBERS

A sense of pattern is crucial in mathematical thinking. Numbers which can be interpreted as geometric figures are ideal for developing that sense, because the many patterns among them involve only positive integers, can be found with little computation, and can often be understood from a geometric point of view. These *figurate* numbers turn up in widely scattered parts of mathematics and have long been a popular recreational topic.

3.1 TRIANGULAR AND SQUARE NUMBERS

The first few triangular numbers are 1, 3, 6, 10, 15, and 21, corresponding to the arrays in Fig. 3.1. The triangular pattern for 10 recalls set-up bowling pins, and that for 15 is the way pool balls are racked up at the start of the game. There is nothing especially triangular about an array of only one dot, but 1 is still counted as a triangular number because it fits the patterns common to all triangular numbers. Of course, the list of triangular numbers can be extended. What are the next few? Try to find them by extending the number patterns in Fig. 3.1, and as a check draw their pictures and count the dots.

We met some higher triangular numbers in Chap. 1, though they were not identified as such, when we found sums like $1 + 2 + 3 + \cdots + 9 + 10$ and $1 + 2 + \cdots + 99 + 100$. Adding all the integers from 1 up through a given number always yields a triangular number, as shown for $1 + 2 + \cdots + 9 + 10$ in Fig. 3.2.

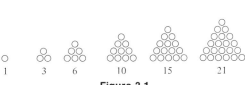

Figure 3.1

Figure 3.2

Find these numbers:
(*a*)   Half of 1 · 2 =
(*b*)   Half of 2 · 3 =
(*c*)   Half of 3 · 4 =
(*d*)   Half of 4 · 5 =
(*e*)   Half of 5 · 6 =

Do you recognize the results? Try to explain them before you read on.

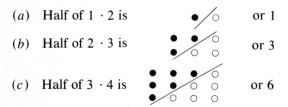

(*a*)   Half of 1 · 2 is   or 1

(*b*)   Half of 2 · 3 is   or 3

(*c*)   Half of 3 · 4 is   or 6

Can you make up similar pictures to explain why half of 4 · 5 and half of 5 · 6 are triangular numbers?

What if we start with square instead of triangular arrays? The three arrays in Fig. 3.3 are of this type, and so the numbers they represent are naturally called *squares*.

What are the next few higher squares? Can you see why when a number is multiplied by itself it is said to be *squared*? Since 1 · 1 = 1, we count 1 as a square number, because it fits the pattern, even though an array of one object has no special shape.

We saw before that sums like 1 + 2 + 3 + 4 and 1 + 2 + 3 are always triangular numbers. What about the analogous sums of odd integers?

(*a*)   1 =
(*b*)   1 + 3 =
(*c*)   1 + 3 + 5 =
(*d*)   1 + 3 + 5 + 7 =
(*e*)   1 + 3 + 5 + 7 + 9 =

Find these and try to explain them before you read on.

Can you see how Fig. 3.4 explains the pattern? Can you make a similar picture to explain why 1 + 3 + 5 + 7 + 9 is a square? Could you continue the process higher still?

| 4 | 9 | 16 | | 1 | 1 + 3 | 1 + 3 + 5 | 1 + 3 + 5 + 7 |

Figure 3.3   **Figure 3.4**

Try adding two consecutive triangular numbers, like this:

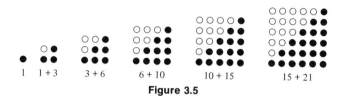

Can you explain this surprising result with pictures? Try it, and then look at Fig. 3.5.

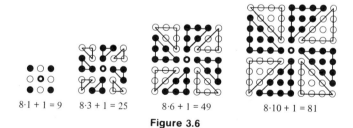

Figure 3.5

Here is another interesting experiment. Multiply any triangular number by 8 and add 1 to the product.

(a) $1 \cdot 8 + 1 =$
(b) $3 \cdot 8 + 1 =$
(c) $6 \cdot 8 + 1 =$
(d) $10 \cdot 8 + 1 =$

What do you notice? Can you explain this pictorially? Think a bit, then look at Fig. 3.6.

$$8 \cdot 1 + 1 = 9 \qquad 8 \cdot 3 + 1 = 25 \qquad 8 \cdot 6 + 1 = 49 \qquad 8 \cdot 10 + 1 = 81$$

Figure 3.6

The most interesting patterns have been saved for the exercises. Compute carefully and be alert for patterns; there are many to be discovered.

Exercises 3.1

1 The first triangular numbers are 1, 3, 6, 10, 15, 21, and 28. Calculate the next eight triangular numbers. (They will be needed in Prob. 3.)

2 (a) The first few square numbers are (counting $0 \cdot 0 = 0$ as a square):

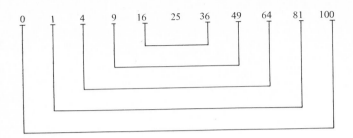

The lines drawn here bring out a certain symmetry in the final digits. Compute the next 10 square numbers and find out whether the pattern continues.

(b) Are there any square numbers which end in 2, 3, 7, or 8?

3 Is there a pattern in the units digits of the triangular numbers similar to that found in Prob. 2 for the square numbers?

4 *Compute*:

(a) $0 \cdot 1 \cdot 2 \cdot 3 + 1$ (b) $1 \cdot 2 \cdot 3 \cdot 4 + 1$ (c) $2 \cdot 3 \cdot 4 \cdot 5 + 1$

What does this seem to have to do with figurate numbers? Try to predict what $3 \cdot 4 \cdot 5 \cdot 6 + 1$ is before computing it. (The relationship discovered here can be proved to hold for the product of *any* four consecutive integers plus 1, but the proof is beyond this book.)

5 (a) Compute:

$14 \cdot 1 + 2 \cdot 3 + 1 =$

$14 \cdot 3 + 2 \cdot 6 + 1 =$

$14 \cdot 6 + 2 \cdot 10 + 1 =$

(b) What kind of numbers are being multiplied by 14 and 2 in each line?

(c) What kind of numbers are the answers?

(d) Try to extend the pattern one more line and check it by doing the arithmetic.

6 A certain club begins each meeting by having each member present shake hands just once with every other. How many handshakes would this involve if the number of members present is:

(a) 2   (b) 3   (c) 4   (d) 5

(e) What do you notice about your answers in parts (a) to (d)?

(f) Suppose the first two members to arrive at the meeting have already shaken hands when a third appears. How many new handshakes are needed? Now suppose a fourth arrives; how many more new handshakes are needed? Continuing this way, find how many handshakes would be needed if the attendance increased one member at a time up to 101.

(g) In Sec. 1.3 we saw that $1 + 2 + 3 + \cdots + 100 = 5{,}050$. We have seen above that this is a triangular number. How is this related to part (f)?

*3.2 EXTENSION TO THREE DIMENSIONS

Triangular and square numbers can be generalized in two ways. One way leads to numbers for figures with more sides but still in the plane, while the other leads to three-dimensional figurate numbers.

The three-dimensional analog of a triangle is a pyramid whose base has three sides, known as a *tetrahedron*. (The prefix *tetra-* comes from the Greek word for "four" and refers here to the fact that a tetrahedron is a solid figure bounded by four triangles, including the base.) The first few tetrahedral numbers are shown in Fig. 3.7. Once again, we count 1 as a tetrahedral number, even though an array of one dot has no special shape.

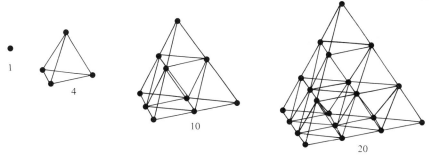

Figure 3.7

The three-dimensional analog of the square is the *cube*. The three smallest cubes are illustrated in Fig. 3.8. The cube with 2 balls on each edge is 8, or $2 \cdot 2 \cdot 2$, and the cube with 3 balls on each edge is 27, or $3 \cdot 3 \cdot 3$. The cube with 4 balls on each edge is 64, or $4 \cdot 4 \cdot 4$, and the next larger cubes are $5 \cdot 5 \cdot 5 = 125$, $6 \cdot 6 \cdot 6 = 216$, and $7 \cdot 7 \cdot 7 = 343$. They are too large to picture easily.

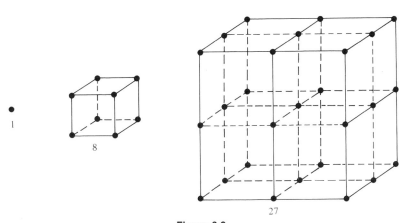

Figure 3.8

Exercises 3.2

1 Compute these sums of the first few triangular numbers.

$$1 = 1$$
$$1 + 3 =$$
$$1 + 3 + 6 =$$
$$1 + 3 + 6 + 10 =$$
$$1 + 3 + 6 + 10 + 15 =$$

Do you recognize the sums? Try to explain them by interpreting them geometrically.

2 The first 11 cubes (counting $0 \cdot 0 \cdot 0 = 0$ as a cube) are $0, 1, 8, 27, 64, 125, 216, 343, 512, 729,$ 1,000. Can you find a relation between their last digits analogous to those found for square and triangular numbers in Exercises 3.1?

3 What have these sums of consecutive odd numbers to do with figurate numbers? (To explain this pattern is not easy without algebra.)

$$1 \qquad =$$
$$3 + 5 \qquad =$$
$$7 + 9 + 11 \qquad =$$
$$13 + 15 + 17 + 19 \qquad =$$
$$21 + 23 + 25 + 27 + 29 =$$

Predict the sum $31 + 33 + 35 + 37 + 39 + 41$ from the pattern, and then check the prediction.

4 A truly remarkable pattern (which is not easy to explain) is in the following sums of cubes. Find it.

$$1 =$$
$$1 + 8 =$$
$$1 + 8 + 27 =$$
$$1 + 8 + 27 + 64 =$$
$$1 + 8 + 27 + 64 + 125 =$$

5 In Sec. 3.1 we saw that if two consecutive positive integers are multiplied and the product divided by 2, the result is always a triangular number. Now multiply three consecutive positive integers and divide by 6 to get another interesting result.

(a) $1 \cdot 2 \cdot 3$ divided by 6 is
(b) $2 \cdot 3 \cdot 4$ divided by 6 is
(c) $3 \cdot 4 \cdot 5$ divided by 6 is
(d) $4 \cdot 5 \cdot 6$ divided by 6 is
(e) $5 \cdot 6 \cdot 7$ divided by 6 is

Do you recognize these?

6 The first eight tetrahedral numbers are $1, 4, 10, 20, 35, 56, 84, 120$. Select any three consecutive numbers in this list, multiply the middle one by 4 and add the other two, as in the examples below.

(a) $1 + 4 \cdot 4 + 10 =$ (b) $4 + 4 \cdot 10 + 20 =$
(c) $10 + 4 \cdot 20 + 35 =$ (d) $20 + 4 \cdot 35 + 56 =$
(e) What do you predict for $35 + 4 \cdot 56 + 84$ and $56 + 4 \cdot 84 + 120$? Check your predictions.

7 There are nine rectangles in Fig. 3.9 if one counts the outer square, the four inner squares, and the four shaded in Fig. 3.10.

(a) How many rectangles are there in the 3 unit square in Fig. 3.11?

Figure 3.9

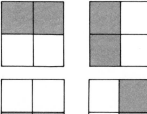

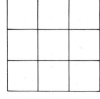

Figure 3.11

Figure 3.10

To count them, count how many are:

1 unit wide and 1 unit long	____	
1 unit wide and 2 units long	____	Total with shortest side 1 _____
1 unit wide and 3 units long	____	
2 units wide and 2 units long	____	Total with shortest side 2 _____
2 units wide and 3 units long	____	
3 units wide and 3 units long	____	Total with shortest side 3 _____
Total	____	

Do you recognize these?

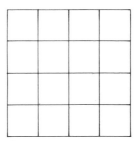

Figure 3.12

(b) How many rectangles are there in a 4 by 4 square like the one in Fig. 3.12?

(c) How does this problem relate to Prob. 4? (The connection is fairly easy to find, but to prove it requires some algebra.)

8 Where are figurate numbers involved in the "Twelve Days of Christmas"?

9 Consider the sums

$$1 = 1$$
$$1 + 4 = 1 + 4$$
$$1 + 4 + 9 = 4 + 10$$
$$1 + 4 + 9 + 16 = 10 + 20$$
$$1 + 4 + 9 + 16 + 25 = 20 + 35$$

(a) What kind of numbers are being added on the left?

(b) What kind of numbers are being added on the right?

(c) Extend the pattern another line and check it.

(d) Figure 3.13 illustrates one of the cases. How?

10 (a) Compute:

$$1 =$$
$$4 =$$
$$1 + 9 =$$
$$4 + 16 =$$
$$1 + 9 + 25 =$$
$$4 + 16 + 36 =$$

(b) What do you notice about the numbers being added and the sums?

(c) Extend the pattern a line or two more, then check it by doing the arithmetic.

★(d) Can you explain this pattern with drawings something like that in Prob. 9?

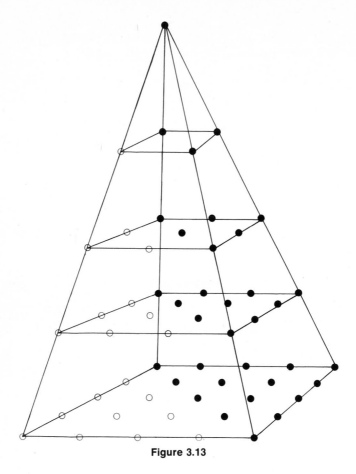

Figure 3.13

11　(*a*) Compute

$$8 \cdot 0 + 1 =$$

$$8 \cdot 1 + 2 =$$

$$8 \cdot 4 + 3 =$$

$$8 \cdot 10 + 4 =$$

$$8 \cdot 20 + 5 =$$

(*b*) What kind of numbers are multiplied by 8 here? What kind of numbers result from these computations?

(*c*) Extend the pattern another line or two, then do the arithmetic to check it.

12　(*a*) Compute

$$4(1 + 4) =$$

$$4(4 + 10) =$$

$$4(10 + 20) =$$

$$4(20 + 35) =$$

(*b*) Try to extend the pattern another line or two and check it.

13 (*a*) Compute

$$0 + 6 \cdot 1 \ + 4 =$$
$$1 + 6 \cdot 4 \ + 10 =$$
$$4 + 6 \cdot 10 + 20 =$$
$$10 + 6 \cdot 20 + 35 =$$

(*b*) Extend the pattern a line or two further, then do the arithmetic to check it.
Problem 8 of Exercises 17.3 is the basis for a geometric explanation of this pattern.

CHAPTER 4
PRIMES AND FACTORING

Prime numbers are fundamental building blocks of the number system and a source of intriguing questions, many unanswered today. Here you will meet prime numbers and learn something of their role in mathematics. If this sounds ambitious, be assured; this chapter involves only positive integers and requires no more computational skill than previous chapters.

4.1 PRIMES

A *multiple* of a number is the result of multiplying it by an integer. For example, 4, 8, 12, . . . are multiples of 4. Figure 4.1 is a table of multiples. The shaded row and column contain the numbers to be multiplied (and incidentally the multiples of 1); other rows and columns contain multiples of other numbers.

The unshaded part of Fig. 4.1 is interesting because some numbers appear in it several times and others not at all. Of course, the reason a number does not appear may be that the table does not go high enough. For example, 50 does not

9	18	27	36	45	54	63	72	81
8	16	24	32	40	48	56	64	72
7	14	21	28	35	42	49	56	63
6	12	18	24	30	36	42	48	54
5	10	15	20	25	30	35	40	45
4	8	12	16	20	24	28	32	36
3	6	9	12	15	18	21	24	27
2	4	6	8	10	12	14	16	18
1	2	3	4	5	6	7	8	9

Figure 4.1

appear in Fig. 4.1, but it would if the table were extended one more row. On the other hand, extending the table will never produce a 7 in the unshaded part. Can you see why? Can you think of other numbers which, like 7, will never appear in the unshaded part of the table?

To be in the unshaded part of Fig. 4.1, a number must be a multiple of positive integers other than itself and 1. Any such number is said to be *composite*. Integers greater than 1 which are not composite are called *primes*; 7 is a prime. No prime can appear in the unshaded part of Fig. 4.1, even if it is extended, because by definition a prime is not a multiple of any positive integer except itself and 1. We shall apply the terms prime and composite only to integers greater than 1, not to other numbers such as fractions or negative integers. In particular, we consider 1 as being neither prime nor composite but in a class of its own.

4.2 FACTORING INTO PRIMES

A number which is expressed as a product is said to be *factored*. An idea of the role of primes can be gained by considering several ways to factor 36. We might begin by factoring 36 as 4 · 9. Both 4 and 9 are composite and can in turn be factored, so that $36 = 2 \cdot 2 \cdot 3 \cdot 3$, as shown schematically in Fig. 4.2. The process ends there, as 2 and 3 are both prime.

What if we begin instead by factoring 36 as 3 · 12? Then we factor the 12 as 2 · 6 and factor the 6 in turn as in Fig. 4.3. The factors in the bottom row in Fig. 4.3 are the same as those found in Fig. 4.2, though in a different order. If you began by factoring 36 as 2 · 18, would you still come out eventually with two factors of 2 and two of 3? What if you began by factoring 36 as 6 · 6? If you begin with another number would a similar result occur? Experiment a bit, then read on.

Any composite number can be factored, and any factors which are composite can in turn be factored. This can be continued until the original number is factored as a product of primes. Further, the prime factorizations of a given composite number differ at most in the order in which the factors are written. (The difference is unimportant, since multiplication is commutative.) We shall not prove this remarkable fact, but we shall use it implicitly by speaking of <u>the</u> *prime factorization* of a positive integer. Because each composite number can

Figure 4.2

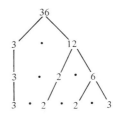

Figure 4.3

be factored into primes in essentially one way, we may regard the primes as the blocks from which composite numbers are assembled.

Exercises 4.2

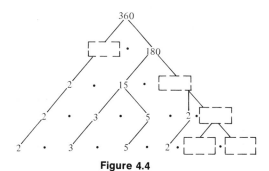

Figure 4.4

1 Fill in the gaps in the factor tree in Fig. 4.4. *Note*: The product in each row is 360.
2 Factor 60 as a product of two factors in as many ways as possible and then refine each factorization to a product of primes. Check that the prime factorizations are all essentially the same.
3 Which of these, if any, are primes? Find the prime factorizations of those which are composite.
   (*a*) 23   (*b*) 25   (*c*) 57   (*d*) 83
   (*e*) 87   (*f*) 91   (*g*) 97   (*h*) 128
*4 Plato, like most ancient Greeks, had a mystical view of numbers. In "The Laws" he states that the ideal city has just 5,040 lots. Find a way to factor 5,040 (*not* into primes) which suggests why that particular number might appeal to a mystic.

4.3 THE SIEVE OF ERATOSTHENES

To find out whether or not a given number is prime can be quite bothersome. In theory the problem is simple; one need only test to see if the given number is a multiple of any positive integer other than 1 or itself. In practice this can involve tedious trial and error. It would be very useful to have a quick test to determine whether a number is prime, but none is known. There is, however, a way to find all the primes less than a given number. This method is known as the sieve of Eratosthenes.[1] We shall use it to find all the primes less than 100. For this you will need pens to write in four colors and a large copy of Fig. 4.5.

Step 1 The smallest prime is 2, but all higher multiples of 2 are composite. Strike them from Fig. 4.5 or your copy of it by putting a horizontal line through each of them, like this: 4 Look for a visual pattern in the numbers you cross

---

[1]Eratosthenes (approximately 274–194 B.C.) was a Greek poet and geographer. After about 240 B.C. he taught at the University of Alexandria, where he was librarian. He computed the circumference of the earth, and if his unit of length was 517 feet, as is generally believed, his value was within 50 miles of modern calculations.

```
 1   2   3   4   5   6   7   8   9  10

11  12  13  14  15  16  17  18  19  20

21  22  23  24  25  26  27  28  29  30

31  32  33  34  35  36  37  38  39  40

41  42  43  44  45  46  47  48  49  50

51  52  53  54  55  56  57  58  59  60

61  62  63  64  65  66  67  68  69  70

71  72  73  74  75  76  77  78  79  80

81  82  83  84  85  86  87  88  89  90

91  92  93  94  95  96  97  98  99 100
```

Figure 4.5

out. The primes less than 100 are among the numbers which are *not* crossed out.

Step 2   The next prime is 3, since it is not a multiple of a smaller prime. Cross off all multiples of 3, except 3 itself, with a line from upper right to lower left like this: 9̷. Use a color different from step 1, and cross off numbers like 6 and 12 at this step again, even though they were already eliminated by step 2. Again there is a visual pattern. To see it, view the page from the lower left-hand corner.

Step 3   The next prime is 5. Cross out all higher multiples of 5 by drawing a line through it from upper left to lower right, like this: 1̷0̷. Again, a new color will bring out the visual pattern.

Step 4   The next prime is 7. Eliminate higher multiples of 7 (but not 7 itself) by enclosing each in a square like this: |14|. This time the visual pattern is more subtle, and a bright new color will help make it stand out of the cluttered array.

How long do we continue? Try crossing off multiples of the next prime, 11. Does this eliminate any numbers not already crossed off in earlier steps? Would crossing off multiples of 13 eliminate any numbers not yet crossed off? If you are not sure, try it.

To bring out what is going on, fill in the table in Fig. 4.6 and find the pattern in it.

Step	Eliminates multiples of	Smallest number crossed out in this step but not in an earlier step
1	2	4
2	3	
3		25
4		

Figure 4.6

Where have you seen the numbers in the right-hand column before? What is the pattern here? If you extended the sieve of Erastothenes to higher numbers, what is the first number 11 would eliminate which had not already been eliminated by a smaller prime? Can you explain the pattern in the table? *Hint*: If a number is a multiple of more than one prime, which prime eliminates it first?

Today there are tables of primes made up by using the sieve of Erastosthenes or modifications of it. Of course with the tables there is no need for the sieve in practical work. It was included here for the insights it yields into the number system. *Save your modified copy of Fig. 4.5 to refer to in Sec. 5.3.*

There is reason to suspect that some of the puzzling aspects of prime numbers (see Sec. 4.5) are related more closely to the fact that they can be generated by a sieving process than to their lack of factors.

Exercises 4.3

1   Use your sieve of Eratosthenes to list all primes less than 100.
2   How many primes are there between:
   (*a*) 0 and 20   (*b*) 20 and 40   (*c*) 40 and 60
   (*d*) 60 and 80   (*e*) 80 and 100
   (*f*) It has been found that primes are rarer among higher numbers. Do these answers bear that out?
3   If you used the sieve of Eratosthenes to find all primes less than 200, what is the largest prime whose multiples you would have to cross out?
4   Are any two consecutive integers both multiples of the same prime? Justify your answer.
5   Each prime is a multiple of just two positive integers, itself and 1. Find at least five numbers which are each multiples of just three positive integers. What property do they have in common?

*4.4 HOW MANY PRIMES ARE THERE?

Primes are not as common among high numbers as among low ones. In fact, as one ascends to very large numbers, the primes become so rare that one may suspect they will eventually die out entirely. Is there some huge number above which there are no more primes?

This question was raised and elegantly answered in Euclid's "Elements," using indirect reasoning, which is best understood in terms of an analogy.[2] Imagine yourself deep in a tunnel, from which you want to escape. One way leads out, while the other is a dead end, but you do not know which is which. You try one way; if it dead-ends, the other way leads out. Naturally, you would like to reach either the exit or a dead end quickly. If you reach neither an exit

[2]Little is known of Euclid's life; even the dates of his birth and death are uncertain. He was probably Athenian, and he taught mathematics at the university of Alexandria. The "Elements," written around 300 B.C., summarized the geometry and number theory of the time, and in logical organization and rigor it was a model of exposition unequaled for over 2,000 years. Since the invention of printing over 1,000 editions of the "Elements" have appeared, second only to the Bible.

nor a dead end, you have no way of telling whether you are going in the right direction. You would welcome anything that indicates you are on a dead end, since then you could confidently go the other way.

The number of primes is either infinite or finite. These are the only two possibilities, and they correspond to directions in the tunnel. We shall start off in one direction, assuming that the number of primes is finite. Then we shall show that that direction leads to a dead end, namely a logical contradiction.

If there is only a finite number of primes, we can multiply all of them together. Their product, which we shall call M, may be too large to calculate, but there must be such a number. Now consider the number $M + 1$. Can $M + 1$ be a multiple of 2? No, because no two consecutive numbers are multiples of 2 (this is clear from the sieve of Eratosthenes). Is $M + 1$ a multiple of 3? Is $M + 1$ a multiple of 5? No, because M is a multiple of each of these, and no two consecutive numbers are both multiples of 3 or of 5. In fact $M + 1$ is not a multiple of any of the primes that were multiplied to produce M. Therefore $M + 1$ is either a prime or a multiple of some prime of which M is not a multiple. Either way, M itself cannot possibly be the product of *all* primes. This is our dead end. We can conclude that there are infinitely many prime numbers.

People who meet indirect reasoning for the first time sometimes feel swindled on a technicality. But although the contradiction is contrived, that is not a flaw; the contradiction is the device that prevents us from wandering down a long dead end. To illustrate the proof concretely, suppose someone thought that the only primes are those already found under 100. Multiply them all together, and call their product $M = 2 \cdot 3 \cdot 5 \cdot 7 \cdot 11 \cdot 13 \cdot 17 \cdot 19 \cdot 23 \cdot 29 \cdot 31 \cdot 37 \cdot 41 \cdot 43 \cdot 47 \cdot 53 \cdot 59 \cdot 61 \cdot 67 \cdot 71 \cdot 73 \cdot 79 \cdot 83 \cdot 89 \cdot 97$. M is so large we do not want to compute it, but that does not matter. Now $M + 1$ is either prime or composite, but it is not a multiple of any of the primes that were multiplied to form M. Therefore $M + 1$ is either a multiple of primes not multiplied to form M or is itself a new prime. Either way, there are primes other than those multiplied to form M.

4.5 PUZZLES ABOUT PRIMES

Part of the fascination of primes is the abundance of questions about them which are so easy to state but so hard to answer. One straightforward-sounding problem is to find a practical formula from which all the primes can be computed, but this seems to be nowhere near solution. A more modest goal would be to find a quick test to distinguish primes from composite numbers, but this has not yet been possible either. The difficulty with problems like these is that the primes are distributed in annoyingly irregular fashion. This irregularity underlies other unsolved problems about primes. For example, it is probably true that there is at least one prime between any two consecutive squares, but so far nobody has been able to prove it. Again, although primes get rarer at a fairly predictable rate as one ascends to higher numbers, one still seems to meet pairs of *twin primes*. These are pairs of consecutive odd integers, such as 11 and 13 or 29 and 31, both of which are primes. The largest pair of twin primes known at present is

$$140,737,488,353,699 \quad \text{and} \quad 140,737,488,353,701,$$

but as larger computers are developed, it seems likely that more will be found. It is widely believed that there are infinitely many pairs of twin primes, but nobody has been able to prove or disprove it.

A different but equally baffling problem was raised by the Swiss mathematics teacher Christian Goldbach, in 1742. He conjectured that every even integer greater than 2 can be expressed as a sum of just two primes. No exceptions are known, but there is always the chance that an exception will be found tomorrow. The prospects for a theoretical proof of Goldbach's conjecture appear very dim.

In addition to the many unanswered questions about primes there are dozens of relationships involving primes which are well understood but curious just the same. You will find some in Sec. 5.3 and Exercises 6.5.

Exercises 4.5

1 Adding just three primes at a time (using the same prime more than once if necessary), we have

$$6 = 2 + 2 + 2$$
$$7 = 2 + 2 + 3$$
$$8 = 2 + 3 + 3$$
$$9 = 3 + 3 + 3$$
$$10 = 2 + 3 + 5$$
$$11 = 3 + 3 + 5$$

(*a*) Can you make other numbers larger than 11 this way? Try it up to 20.
*(*b*) Do you think every integer greater than 6 can be expressed this way? (Would that be so if Goldbach's conjecture were true?)

2 (*a*) The prime 83 is the sum of the squares of three different primes. What are they?
(*b*) Find another prime less than 200 which is the sum of the squares of three different primes. (It is widely believed, though it has never been proved, that there are infinitely many primes which are the sum of the squares of three different primes.)

CHAPTER 5
EXPONENTS

Here you will be introduced to exponents. You will use them to abbreviate repeated multiplication much the way multiplication itself can be used to abbreviate repeated addition. This chapter, like those before, deals only with whole numbers. Later, especially in Chap. 12, you will see other uses for exponents and generalize the ideas in this chapter.

5.1 EXPONENTS AND REPEATED MULTIPLICATION

In Chap. 3 you met the square numbers $1 \cdot 1$, $2 \cdot 2$, $3 \cdot 3$, $4 \cdot 4$, . . . and the cube numbers $1 \cdot 1 \cdot 1$, $2 \cdot 2 \cdot 2$, $3 \cdot 3 \cdot 3$, $4 \cdot 4 \cdot 4$, ... These are usually written in more compact form like this:

Squares: $1^2, 2^2, 3^2, 4^2, 5^2, \ldots$
Cubes: $1^3, 2^3, 3^3, 4^3, 5^3, \ldots$

Here the raised 2s for the squares and 3s for the cubes are called *exponents*. They indicate how many factors of the given number are to be multiplied. Exponents larger than 3 work the same way. For example, 7^5 means $7 \cdot 7 \cdot 7 \cdot 7 \cdot 7$ and is read "seven to the fifth power" or, for short, "seven to the fifth." The second and third powers of a number are called its *square* and *cube* respectively, even when they cannot be visualized as they were in Chap. 3. For example, $(^-4)^2$ is called the square of $^-4$.

An exponent applies only to the number by which it is written unless grouping symbols explicitly indicate otherwise. For example, $1 + 2^3$ is $1 + 2 \cdot 2 \cdot 2 = 1 + 8$, whereas $(1 + 2)^3$ is $3^3 = 3 \cdot 3 \cdot 3 = 27$. Confusion between the two is especially likely in speech, for the phrase "1 plus 2 cubed" is ambiguous. It is better to be extra careful and say either "1 plus the cube of 2" or "the cube of the sum of 1 and 2," as the case may be.

Exponents are especially handy for dealing with very large and very small numbers. In Chap. 12 you will see how these numbers are dealt with in scientific notation, which is based on powers of 10. For now, however, consider the prime factorization of 10,368, namely $2 \cdot 2 \cdot 2 \cdot 2 \cdot 2 \cdot 2 \cdot 2 \cdot 3 \cdot 3 \cdot 3 \cdot 3$. This expression is so clumsy that one must count the factors even to copy it. How much simpler it is to write $2^7 3^4$.

Exercises 5.1

1 Write in exponential form.
   (a) $3 \cdot 3 \cdot 3 \cdot 3 \cdot 3$ (b) $17 \cdot 17 \cdot 17 \cdot 17 \cdot 17 \cdot 17 \cdot 17 \cdot 17$
   (c) $(^-2)(^-2)(^-2)(^-2)(^-2)(^-2)$ (d) $5 \cdot 5 \cdot 5 \cdot 5 \cdot 5 \cdot 12 \cdot 12 \cdot 12$
   (e) $3 \cdot 7 \cdot 2 \cdot 3 \cdot 11 \cdot 7 \cdot 3 \cdot 2$

2 Evaluate:
   (a) 3^4 and 4^3 (b) 2^5 and 5^2
   (c) What do these examples show?

3 Evaluate:
   (a) 1^3 (b) 1^5 (c) 1^{10} (d) 1^{97} (e) generalize

4 Evaluate:
   (a) 5^1 (b) 14^1 (c) 927^1 (d) generalize

5 Evaluate:
   (a) 10^1 (b) 10^2 (c) 10^3 (d) 10^4 (e) generalize

6 Folding a sheet of paper in half doubles its thickness.
   (a) If the folded paper is folded in half a second time, how many sheets thick is it?
   (b) If the paper is folded in half a third time, how many sheets thick is it?
   (c) How many sheets thick would the paper be after five such folds?
   (d) Why is it a safe bet that nobody can fold an ordinary piece of paper in half nine times with his bare hands?

7 Certain numbers can be expressed as a sum of two or more consecutive positive integers. For example, 3 can be expressed as $1 + 2$, 6 as $1 + 2 + 3$, and 18 as $3 + 4 + 5 + 6$.
   (a) Find the four integers greater than 1 but less than 30 which *cannot* be so expressed.
   (b) What property involving exponents do these four numbers have in common?

8 (a) Check to see that these curious relations are true:

$$1 + 2 = 3$$

$$4 + 5 + 6 = 7 + 8$$

$$9 + 10 + 11 + 12 = 13 + 14 + 15$$

$$16 + 17 + 18 + 19 + 20 = 21 + 22 + 23 + 24$$

   (b) Look at the left-hand number of each line. Do you recognize them?
   (c) Take half of the number in each line just left of the equals sign. What do you notice?
   (d) Can you extend the pattern another line? Check it!

9 (a) Check to see that these curious relations are true:

$$3^2 + 4^2 = 5^2$$

$$10^2 + 11^2 + 12^2 = 13^2 + 14^2$$

$$21^2 + 22^2 + 23^2 + 24^2 = 25^2 + 26^2 + 27^2$$

   (b) Look at the left-hand number in each line. Do you recognize them?
   (c) The numbers whose squares appear just to the left of the equals sign are interesting. Do you recognize them? *Hint*: Each is 4 times a familiar number.
   (d) Extend the pattern one more line and check it.

10 G. H. Hardy, visiting his ailing friend Srinivasa Ramanujan, mentioned that he had ridden in cab number 1,729, a number which struck him as uninteresting. (Only mathematicians could have this conversation!) Ramanujan, a mystic who regarded numbers as his friends, startled Hardy by replying unhesitatingly that, on the contrary, 1,729 is very interesting, as it is the smallest number which can be expressed as the sum of two cubes in two different ways. How?

11 (a) Evaluate $1 + 2^3 + 3^2$.

(b) Insert parentheses in $1 + 2^3 + 3^2$ to change its value to 122.

*(c) Insert two sets of parentheses in $1 + 2^3 + 3^2$ to make its value 900.

12 This array, called *Pascal's triangle*, comes up in many parts of mathematics:[1]

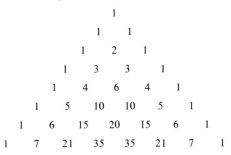

(The triangle can be extended further, as we shall see.) These exercises will bring out some of the marvelous patterns in this triangle.

(a) Add

$$1 =$$
$$1 + 1 =$$
$$1 + 2 + 1 =$$
$$1 + 3 + 3 + 1 =$$
$$1 + 4 + 6 + 4 + 1 =$$
$$1 + 5 + 10 + 10 + 5 + 1 =$$

What do you predict for $1 + 6 + 15 + 20 + 15 + 6 + 1$? Check it!

(b)

$$1 + {}^-1 =$$
$$1 + {}^-2 + 1 =$$
$$1 + {}^-3 + 3 + {}^-1 =$$
$$1 + {}^-4 + 6 + {}^-4 + 1 =$$

What do you predict for $1 + {}^-5 + 10 + {}^-10 + 5 + {}^-1$? Check it!

(c) Find the triangular numbers in Pascal's triangle.

(d) Find the tetrahedral numbers in Pascal's triangle.

*(e) Figure 5.1 is a street map. From each intersection you must move down, but you may choose whether to go to the right or the left. How many paths are there from the start to each lettered intersection? What do you notice?

(f) Figure out how to extend Pascal's triangle two or more rows by finding patterns in it. The patterns you found before should all continue.

---

[1]Blaise Pascal (1623–1662) was born in Clermont, France. Against his father's wishes he studied mathematics and physics, and by sixteen he had already found some deep new results in geometry. A few years later he invented a mechanical calculator. At twenty-seven he put aside these studies and turned to theology. The little mathematics he did thereafter included his "Treatise of the Arithmetic Triangle," for which Pascal's triangle is named.

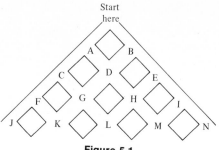

Start
here

A   B
C   D
F   G   H   E
J   K   L   M   N   I

Figure 5.1

(*g*)

$$1^2 + 1^2 =$$
$$1^2 + 2^2 + 1^2 =$$ Each of these answers
$$1^2 + 3^2 + 3^2 + 1^2 =$$ appears in the triangle
$$1^2 + 4^2 + 6^2 + 4^2 + 1^2 =$$

Use this pattern to predict $1^2 + 5^2 + 10^2 + 10^2 + 5^2 + 1^2$. Then extend the triangle far enough to check your prediction.

13   The numbers 1, 1, 2, 3, 5, 8, 13, 21, 34, . . . , where each number in the list is the sum of the two before, are called *Fibonacci numbers* after their discoverer.[2] So curious are the many relationships involving these numbers that a mathematical journal, the *Fibonacci Quarterly*, is devoted to them and related topics. These exercises bring out some of their properties.
(*a*) Squares of successive Fibonacci numbers:

$1^2 + 1^2 =$ _____ $1^2 + 2^2 =$ _____ $2^2 + 3^2 =$ _____
$3^2 + 5^2 =$ _____ $5^2 + 8^2 =$ _____ $8^2 + 13^2 =$ _____

Can you predict $13^2 + 21^2$ and $21^2 + 34^2$? Try, then check your predictions by computing.
(*b*) What pattern do you see here?

$$1^2 + 1^2 = 1 \cdot 2$$
$$1^2 + 1^2 + 2^2 = 2 \cdot 3$$
$$1^2 + 1^2 + 2^2 + 3^2 = 3 \cdot 5$$
$$1^2 + 1^2 + 2^2 + 3^2 + 5^2 = 5 \cdot 8$$

Extend the pattern two more lines. Does the arithmetic still check?
(*c*) What pattern do you see here?

$1 \cdot 3 = 2^2 - 1$
$2 \cdot 5 = 3^2 + 1$
$3 \cdot 8 = 5^2 - 1$
$5 \cdot 13 = 8^2 + 1$
$8 \cdot 21 = 13^2 - 1$

Extend the pattern to two more cases, then check it by doing the arithmetic.
(*d*) We saw some patterns in Pascal's triangle in Problem 12. Now add along the arrows in Fig. 5.2 to find another pattern.

[2]See note on Fibonacci on page 96.

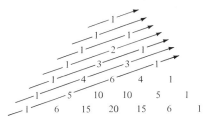

Figure 5.2

14 Versions of this puzzle have been traced to ancient Egypt.

> As I was going to St. Ives,
> I met a man with seven wives,
> Each wife had seven sacks,
> Each sack had seven cats,
> Each cat had seven kits:
> Kits, cats, sacks, and wives
> How many were going to St. Ives?

Neglecting the trick answer, how many people, cats, kits, and sacks did the traveler meet?

15

$2 = 1^2 + 1^2$	$3 = 1^2 + 1^2 + 1^2$	$4 = 2^2$
$5 = 1^2 + 2^2$	$6 = 1^2 + 1^2 + 2^2$	$7 = 1^2 + 1^2 + 1^2 + 2^2$	$8 = 2^2 + 2^2$
$9 = 3^2$	$10 = 1^2 + 3^2$	$11 = 1^2 + 1^2 + 3^2$	$12 = 2^2 + 2^2 + 2^2$

Here each number is expressed as a sum of as few squares as possible.
(*a*) Continue this, at least up to 30.
(*b*) What is the largest number of squares you needed in part (*a*) to make any one number?

(Every positive integer can be expressed as a sum of at most four squares. There are similar results for cubes and higher power powers, but this area of mathematics has many unsolved problems.)

5.2 LEAST COMMON MULTIPLES

Suppose you are given two or more positive integers. What is the smallest number which is a multiple of each and every one of them? This number is called the *least common multiple* of the given numbers, and understanding how to find it is an important step in learning how to add fractions. We shall not add fractions until Chap. 7, but finding least common multiples is appropriate here as an illustrative use of factors and exponents.

To get an idea of the situation, consult Fig. 4.5 as you modified it for the sieve of Eratosthenes. Suppose a number is a common multiple of 2 and 3; that is, it is a multiple of 2 and a multiple of 3. What marks did you put on it? How did you mark all common multiples of 2 and 5? How did you mark all common multiples of 3 and 5? Experiment a bit. Find the least common multiples of several pairs of these primes such as 2 and 7 or 3 and 5. Do you see a pattern? Does the pattern hold if the numbers are not all primes? If not, can it be modified to fit that case? Do not read on until you have experimented enough to get a feel for the situation. Only then will what follows make sense to you.

You probably found that the least common multiple of several primes is their product but that if the given numbers are not primes, their product is not necessarily their *least* common multiple. An example of this is 4 and 6; their product is 24, but their least common multiple is 12. This is closely related to the question of prime factorization, as the following example shows.

Suppose we want to find the least common multiple of 18 and 21. The number we seek is a multiple of 21, and so its prime factorization must include factors of 3 and 7. Likewise it is a multiple of 18, and so its prime factorization must include the prime factorization $2 \cdot 3 \cdot 3$ for 18. The number $2 \cdot 3 \cdot 3 \cdot 7$ meets both conditions. Do you see why? [*Hint*: Group the factors as $(2 \cdot 3 \cdot 3) \cdot 7$ and $(2 \cdot 3)(3 \cdot 7)$. What factors appear in the parentheses?] Is any smaller number a common multiple of 18 and 21?

The idea is the same when we begin with more than two numbers. For example, can you find the least common multiple of 12, 15, 16, and 18? Try it on your own, and then check yourself by reading on.

Factoring into primes $12 = 2 \cdot 2 \cdot 3$, $15 = 3 \cdot 5$, $16 = 2 \cdot 2 \cdot 2 \cdot 2$, and $18 = 2 \cdot 3 \cdot 3$. The least common multiple must have four factors of 2 to be a multiple of 16. It must have two factors of 3 to be a multiple of 18, and it must also have a factor of 5 to be a multiple of 15. It is therefore $2 \cdot 2 \cdot 2 \cdot 2 \cdot 3 \cdot 3 \cdot 5 = 720$. Can you (see Prob. 1) insert parentheses to show that $2 \cdot 2 \cdot 2 \cdot 2 \cdot 3 \cdot 3 \cdot 5$ is indeed a common multiple of 12, 15, 16, and 18?

Exponents can shorten and clarify the work in some cases. For example, to find the least common multiple of 108 and 288, we might express their prime factorizations by means of exponents:

$$108 = 2^2 \cdot 3^2 \qquad \text{and} \qquad 288 = 2^5 \cdot 3^2$$

From this it is possible to see directly that the least common multiple of 108 and 288 is $2^5 \cdot 3^3$. Can you see why?

Exercises 5.2

1　Insert parentheses into $2 \cdot 2 \cdot 2 \cdot 2 \cdot 3 \cdot 3 \cdot 5$ to show it is a multiple of:
(*a*) 12　　(*b*) 15　　(*c*) 16　　(*d*) 18
2　Use the prime factorization of 385, which is $5 \cdot 7 \cdot 11$, to answer these questions *without* actually dividing:
(*a*) By what must 5 be multiplied to get 385?
(*b*) By what must 7 be multiplied to get 385?
(*c*) By what must 11 be multiplied to get 385?
(*d*) By what must 35 be multiplied to get 385?
(*e*) By what must 55 be multiplied to get 385?
(*f*) By what must 77 be multiplied to get 385?
3　The prime factorization of 900 is $2^2 \cdot 3^2 \cdot 5^2$. Use that to answer these questions *without* actually dividing:
(*a*) By what must 4 be multiplied to get 900?
(*b*) By what must 60 be multiplied to get 900?
(*c*) By what must 75 be multiplied to get 900?
4　In each case find the least common multiple of the given numbers. You may be able to do some entirely in your head, but the last ones should challenge you.
(*a*) 14, 16　　(*b*) 1, 2, 3, 4,　　　(*c*) 15, 20

(d) 16, 18   (e) 54, 72   (f) 84, 91
(g) 25, 35   (h) 96, 100   (i) 4, 5, 6, 7, 8
(j) 49, 50   (k) 102, 144, 150   (l) 91, 121, 308, 1,001

5 In each case find the least common multiple of the given numbers. Leave your answers in exponential forms like those given.

(a) $2^4 \cdot 7^2$, $2^1 \cdot 7^3$ (b) $3^7 \cdot 11^3$, $3^5 \cdot 11^5$
(c) $2^3 \cdot 3^1$, $5^4 \cdot 11^2$, $5^7 \cdot 11^4$, $3^4 \cdot 5^6$

*5.3 PERFECT NUMBERS

> Six is a perfect number in itself, and not because God created all things in six days; rather the inverse is true; God created all things in six days because this number is perfect, and it would remain perfect even if the work of six days did not exist.

Ancient mystics considered a number to be *perfect* if it is the sum of all positive integers of which it is a multiple, except for itself. Thus 6 is perfect since it is a multiple of 1, 2, and 3, and these add to 6 itself, but 8 is not perfect, since except for itself it is a multiple of 1, 2, and 4 and these total only 7.

Perfect numbers were first studied by mystics (in his twelfth-century book, "Healing of Souls," Rabbi Josef ben Jehuda Ankin recommends their study), but today they are known primarily for the fascinating mathematical problems they lead to and for their colorful history.

Perfect numbers are rare; the eight smallest are

6

28

496

8,128

33,550,336

8,589,869,056

137,438,691,328

2,305,843,008,139,952,128

All these were listed by the French amateur mathematician Mersenne in his "Cogitata" in 1644.[3] He also listed what he claimed to be the next three higher perfect numbers, but two of them have since been shown not to be perfect.

All 24 perfect numbers known to date conform to the elegant formula $2^{(n-1)}(2^n - 1)$, which was known even in Euclid's time. Curiously, however, this formula yields perfect numbers only for certain values of n; $2^{(n-1)}(2^n - 1)$

[3]Marin Mersenne (1588–1648) was a Minim friar who was also an excellent mathematician and scientist. He met regularly with the leading French thinkers of his time, including Descartes (see page 84) and Fermat (see page 147).

is perfect when n is 2, 3, 5, or 7 but not when n is 4, 6, 8, 9 or 10. This leads one to guess that $2^{(n-1)}(2^n - 1)$ is a perfect number whenever n is prime, but things are not so simple, for $2^{(n-1)}(2^n - 1)$ is not perfect when $n = 11$. It turns out that Euclid's formula yields a perfect number only when $2^n - 1$ is itself prime, which it never is when n is composite and only sometimes is when n is prime. The only primes n less than 20,000 for which $2^n - 1$ is prime are

2	3	5	7	13
17	19	31	61	89
107	127	521	607	1,279
2,203	2,281	3,217	4,253	4,423
9,689	9,941	11,213	19,937	

The prime $2^{11,213} - 1$ is almost unimaginably large. If you counted at a rate of 1 million numbers per second, starting 10 billion years ago (which is generally thought to be before the earth was formed) you would not yet have reached $2^{11,213} - 1$. The size of $2^{19,937} - 1$, which was proved prime in 1970, is totally beyond comprehension; to write it in the usual decimal notation would require some 6,000 digits. Only the combined use of sophisticated mathematics and fast computers has made it possible to deal with such large numbers. One must marvel that Mersenne, whose most modern computing device was a quill pen, was able to find, correctly, that $2^{127} - 1$ is prime, for this number, in decimal form, is

$$170,141,183,460,469,231,731,687,303,715,884,105,727$$

Possibly he and his even more brilliant friend Fermat had insights which are no longer known.

Euclid's formula yields all the even perfect numbers, but it says nothing about the possibility of any odd numbers being perfect. This remains an open question. Before you try to win fame by being the first to find an odd perfect number, you should know that if there are any, they have at least 37 digits.

Exercises 5.3

1   Evaluate $2^{(n-1)}(2^n - 1)$ for $n = 2, 3, 4, 5$, and 6. Then verify that the result is a perfect number for $n = 2, 3$, and 5 but not for $n = 4$ or 6.
2   The following sums of odd cubes look rather like some of the sums in Chap. 3, but the relation they reveal is even more beautiful and surprising than those discovered there.
   (a) $1^3 + 3^3$
   (b) $1^3 + 3^3 + 5^3 + 7^3$
   (c) $1^3 + 3^3 + 5^3 + 7^3 + 9^3 + 11^3 + 13^3 + 15^3$
   (d) Try to generalize the pattern (to prove that it always holds is beyond the scope of this book).

3 A number is called *multiply perfect* if the sum of its factors other than itself is an exact multiple of it. (For perfect numbers this multiple is 1.) Other than perfect numbers themselves, many multiply perfect numbers are known, the two smallest being 120 and 672. Verify that these are indeed multiply perfect.

4 (*a*) The numbers 1,210 and 1,184 have a property related to perfect numbers; the sum of the factors of each (except for itself) is the other, and, as such, 1,210 and 1,184 are an example of an *amicable pair*. Verify this.

(*b*) A Bible scholar has suggested that Jacob's present of 220 goats to Esau (Genesis 32:14) is especially significant, because 220 is one member of an amicable pair of numbers. Verify this and find the other member of the pair.

CHAPTER 6
SUBTRACTION
AND DIVISION

So far we have avoided subtraction and division in order to concentrate on addition and multiplication. Now you will see it is not difficult to fill in the gap, as subtraction and division are easier to understand in relation to addition and multiplication than as totally new processes.

6.1 WHAT IS SUBTRACTION?

Suppose you pay for a 17-cent item with a 50-cent piece. Making the change, the clerk says something like this: "Seventeen [*handing you the package*] and three [*handing you three pennies*] makes twenty, and five [*handing you a nickel*] and twenty-five [*handing you a quarter*] is fifty." By beginning at 17 and counting up to 50, he has found the number which must be added to 17 to get 50. That number, called the *difference* of 17 from 50, is written $50 - 17$. The process of computing a difference is called *subtraction*. In our example, the fact that $17 + (3 + 5 + 25) = 50$ shows that $50 - 17 = 33$. Subtraction, then, deals with the question of what must be added to one number to get another. As long as you keep this clearly in mind, you will not find subtraction difficult.

We shall avoid the phrase "take away" in discussing subtraction, for while it clarifies some cases, it obscures others. "If from five cookies in a jar, three are taken away, how many are left?" is a typical subtraction problem stated in terms of taking away, and it is easy to visualize. On the other hand, nobody can take five cookies from a jar with only three in it, and teachers who view subtraction solely in terms of taking away have told many youngsters that 5 cannot be subtracted from 3. Later, when negative numbers are introduced, the students must unlearn this; to subtract 5 from 3 is to find the number which must be added to 5 to get 3, namely ¯2. Logically, one could avoid this difficulty by introducing negative numbers before subtraction, but this is difficult pedagogically, especially since most elementary mathematics texts are not designed that way. The least one can do, however, is avoid tying subtraction exclusively to taking away.

Those who learned subtraction in terms of taking away find certain subtractions, like $3 - {}^-7$, especially confusing. Elaborate explanations have been concocted to justify the fact that the result of this subtraction, 10, is greater than 3. One of the favorites is that subtracting $^-7$ is like losing a debt of \$7, which improves one's fortunes, but this is needlessly farfetched. There is nothing to explain if you bear in mind what subtraction is: $3 - {}^-7$ is the number which must be added to $^-7$ to get 3, and since $^-7 + 10 = 3$, we see that $3 - {}^-7 = 10$.

Exercises 6.1

1 Subtract as indicated, keeping in mind the basic question of what must be added. Use the patterns to check your answers.

(a)	(b)	(c)
$8 - 3$	$19 - 16$	$5 - 2$
$8 - 2$	$19 - 17$	$4 - 2$
$8 - 1$	$19 - 18$	$3 - 2$
$8 - 0$	$19 - 19$	$2 - 2$
$8 - {}^-1$	$19 - 20$	$1 - 2$
$8 - {}^-2$	$19 - 21$	$0 - 2$
$8 - {}^-3$	$19 - 22$	$^-1 - 2$
		$^-2 - 2$

2 Can you reconstruct these addition box puzzles?

(a)

	9	13
12		25
	22	

(b)

	5	
		14
11		27

(c)

		21
	13	
18		50

(d)*

	22	57
	13	41

3 The adder of Sec. 2.4 can be used for subtracting. Can you figure out how? Check that the results it yields are what you would expect, even when negative numbers are involved.

4 Maps give the elevation of the floor of Death Valley as $^-282$ feet, meaning that it is 282 feet below mean sea level. From the floor of Death Valley one can see the top of Mt. Whitney, 14,495 feet above mean sea level.

    (a) How much higher is the top of Mt. Whitney than the floor of Death Valley?

    (b) What subtraction does this illustrate?

5 On first down a quarterback went back to pass but was thrown for a 17-yard loss. How far did he have to go for a first down? What subtraction problem does this illustrate?

6 One wintry evening in a small town in Alaska the temperature fell to $^-47°$F. However, by the following noon it had soared to a balmy $^-27°$F. How much had it risen? What subtraction does this illustrate?

6.2 ADDING THE OPPOSITE

Study these examples, which come in pairs. What do you notice?

Subtraction	Addition
$5 - 3 = 2$	$5 + {}^-3 = 2$
$^-7 - 5 = {}^-12$	$^-7 + {}^-5 = {}^-12$
$100 - 73 = 27$	$100 + {}^-73 = 27$
$15 - {}^-5 = 20$	$15 + 5 = 20$
$^-13 - {}^-7 = {}^-6$	$^-13 + 7 = {}^-6$

Try to generalize from these before you read on.

In each case above, subtracting a number has the same effect as adding its opposite. This is always so; subtracting c from b has the same result as adding its opposite, ^-c, to b. More concisely, $b - c$ is the same number as $b + {}^-c$. It is therefore possible to define subtraction as addition of the opposite. This is simple and direct, but it is inconvenient for use in elementary schools. Subtraction defined that way does not make sense to students who have not already learned to add both positive and negative numbers. The view of subtraction as finding what must be added to one number to get another is logically sound and relatively easy to teach youngsters, but if you find it convenient to subtract by adding opposites, do not hesitate to do so in your own computations.

So far we have used an ordinary minus sign to denote the operation of subtraction, but we have used a raised minus sign to denote a conceptually different operation, taking the opposite of a number. After Section 6.2, however, we shall use an ordinary minus sign for both these operations. This simplifies writing, especially where fractions are concerned, and causes no confusion since subtracting a number has the same effect as adding its opposite.

Exercises 6.2

1 Write an addition with the same answer as each of these, as discussed above:

    (*a*) $7 - 1$ (*b*) $14 - 9$ (*c*) $15 - 15$ (*d*) $2 - 9$

    (*e*) $0 - 18$ (*f*) $3 - 11$ (*g*) $^-12 - 13$ (*h*) $14 - {}^-2$

    (*i*) $123 - {}^-38$ (*j*) $^-4 - {}^-8$ (*k*) $11 - {}^-8$

2 Find the differences in Prob. 1 and check them by adding.

3 (*a*) Is subtraction commutative? Justify your answer.

    (*b*) Is subtraction associative? Justify your answer.

4 Difference patterns. Each row is formed by subtractions in the row above. Work carefully and be alert for patterns. Some blanks are filled in as clues and checks for you. Fill in the rest.

(a) Squares

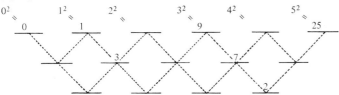

(b) Cubes

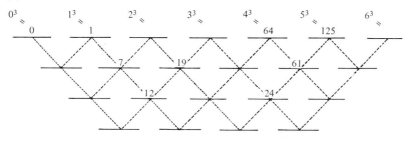

(c) Do the patterns continue when powers of negative numbers are involved? Try it for both squares and cubes.

(d) On the basis of the patterns in parts (a) and (b), what do you predict about the difference pattern for fourth powers? Test your predictions here:

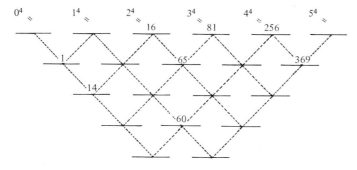

(e) Do you see an overall pattern? On the basis of parts (a), (b), and (d), what do you predict for fifth and sixth powers?

5 Compute and compare these pairs of problems:

(a) $7 - (4 - 1) =$ and $7 - 4 + 1 =$
(b) $5 - (9 - 2) =$ and $5 - 9 + 2 =$
(c) $^-12 - (8 - 3) =$ and $^-12 - 8 + 3 =$
(d) $15 - (^-4 - 5) =$ and $15 + 4 + 5 =$
(e) Generalize. *Hint*: Compare with Prob. 7 of Exercises 2.3.

6.3 METHODS OF SUBTRACTION

Integers are usually subtracted columnwise. For example, to subtract 12 from 97, one subtracts the 2 from the 7 to get 5 and the 1 from the 9 to get 8, so

$97 - 12 = 85$. What have the columnwise subtractions to do with the original problem? The connection goes back to the fact that the 6 in 67 stands for 60 and the 3 in 32 stands for 30, so that the process is

$$97 = 90 + 7$$
$$12 = 10 + 2$$
$$\overline{85 = 80 + 5}$$

What if we tried this with a problem like $52 - 37$? The work would look like this:
$$52 = 50 + 2$$
$$37 = 30 + 7$$
$$\overline{20 + (-5) = 15}$$

Here the answer is correct, but the use of negative numbers is an unnecessary complication. It can be avoided by writing 52 as $40 + 12$ instead of $50 + 2$. Then the work looks this way:

$$52 = 40 + 12$$
$$37 = 30 + 7$$
$$\overline{10 + 5 = 15}$$

Rewriting $50 + 2$ as $40 + 12$ is called *regrouping* or *borrowing*. It is not a difficult process, but it seems to cause some beginners to lose sight of the main question of what must be added. This adds to the distressingly large number of students who forego trying to understand and settle for rote "learning." It leads them away from the view that mathematics is basically common sense and gives them the feeling that it is something to be accepted on faith and parroted back on demand.

A conceptually simpler method for subtraction is the counting technique described earlier. For example, to subtract 389 from 1,227, we proceed this way. Adding 11 to 389 gives us 400, and 600 more yields 1,000. Another 227 gets us to 1,227, so that $1,227 - 389 = 11 + 600 + 227 = 838$. This avoids regrouping and keeps attention on the main question of what must be added. For large numbers it is slow, but it is a good alternative for those who have trouble with the columnwise method, especially since calculators have made emphasis on speed in hand calculation obsolete.

Exercises 6.3

1 Here regrouping is used to make columnwise subtraction easier, but blanks have been left. Fill them in.

(a) $93 = \underline{} + 3 = \underline{} + 13$ (b) $77 = 70 + \underline{} = \underline{} + 17$

$57 = 50 + 7 = 50 + 7$ $39 = \underline{} + 9 = 30 + \underline{}$

$\overline{30 + \underline{} = 36}$ $\overline{30 + \underline{} = \underline{}}$

(c) $235 = 200 + \underline{\hspace{1cm}} + 5 = 100 + \underline{\hspace{1cm}} + 5$

$182 = \underline{\hspace{1cm}} + 80 + 2 = 100 + \underline{\hspace{1cm}} + 2$

$\phantom{(c)\ 235 = 200 + \underline{\hspace{1cm}} + 5 =}\ 50 + \underline{\hspace{1cm}} = \underline{\hspace{1cm}}$

(d) $472 = \underline{\hspace{1cm}} + 70 + 2 = 300 + \underline{\hspace{1cm}} + 2 = 300 + \underline{\hspace{1cm}} + 12$

$195 = 100 + \underline{\hspace{1cm}} + 5 = 100 + 90 + \underline{\hspace{1cm}} = \underline{\hspace{1cm}} + 90 + \underline{\hspace{1cm}}$

$\phantom{(d)\ 195 = 100 + \underline{\hspace{1cm}} + 5 = 100 + 90 +}\ \underline{\hspace{1cm}} + 70 + \underline{\hspace{1cm}} = \underline{\hspace{1cm}}$

2 Try these subtractions by the counting technique. They are arranged in two sequences.

(a) $150 - 147$ (e) $790 - 782$

(b) $200 - 147$ (f) $800 - 782$

(c) $1{,}000 - 147$ (g) $1{,}000 - 782$

(d) $1{,}321 - 147$ (h) $1{,}064 - 782$

3 These addition box puzzles were partly erased. Complete them.

(a)

	-8	17
4		-9

(b)

	-42	
		107
	-8	-14

(c)

		-98
	252	
0		-107

(d)

	-58	102
	41	-137

6.4 WHAT IS DIVISION?

If 8 children share 24 cookies equally, how many will each get? The problem here is to find a number by which 8 can be multiplied to get 24. The missing factor is called the *quotient* of 24 divided by 8, written $24 \div 8$ or 24/8. The process of computing a quotient is called *division*. Sharing problems like the above are good examples of division, but some divisions, such as $-100 \div (-25)$ do not fit that interpretation, just as some subtractions are not easily understood in terms of taking away. As with subtraction, you will find division straightforward if you keep attention on the main problem; in this case it is to find a missing factor.

   One class of division problems cannot be dealt with at all, namely those problems where the known factor (sometimes called the *divisor*) is 0. If, for example, the division $5 \div 0$ could be carried out, the quotient would be a number which multiplied by 0 yields 5. But there is no such number, since 0 times any number is 0. Similarly, no number could be the quotient for any division $x \div 0$, where x is not 0. On the other hand, $0 \div 0$ could be any number at all. It could be 3, since $0 = 3 \cdot 0$, but it could equally well be 1,000, since $0 = 1{,}000 \cdot 0$.

Because there is no sensible way to define division by 0, we avoid it; hence the mathematician's time-honored eleventh commandment: "Thou shalt not divide by 0."

Exercises 6.4

1 Sally had done all the box puzzles on her homework, but her little sister erased most of them. Can you help her?

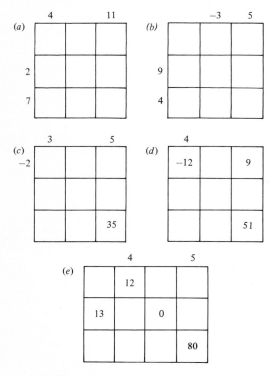

2 Is division associative or commutative? Justify your answer.
3 Suppose $a \div b = c$. Use the meaning of division to complete these statements:
    (a) If a and b are both positive, c is _____.
    (b) If a is positive and b is negative, c is _____.
    (c) If a is negative and b is positive, c is _____.
    (d) If a and b are both negative, c is _____.

6.5 THE DIVISION ALGORITHM

For all but the simplest cases, a systematic procedure, or *algorithm*,[1] for division is helpful. This section shows how such an algorithm, which used to be known as "long division," works. To illustrate, we divide 1,554 by 42; this is a typical division problem, though not especially difficult.

[1]A Latin translation of a ninth-century book on Hindu numeration, by al-Khowarizmi, begins, "Spoken has Algoritmi. . . ." From this comes our word *algorithm* for a systematic calculating procedure.

$$\begin{array}{r} 37 \\ 42\overline{)1554} \\ 126 \\ \hline 294 \\ 294 \\ \hline 0 \end{array}$$

Figure 6.1

Until recently, students were taught to write $42\overline{)1554}$ and then say something like this in working Figure 6.1. "No multiple of 42 is less than 1 or 15, but $3 \cdot 42 = 126$ is less than 155, so write 3 above the second 5 and 126 under 155. Subtracting 126 from 155 leaves 29, and bringing down the 4 makes it 294. By trial we find $42 \cdot 7 = 294$, so we write 7 above the 4, and another 294 under the one we have. Since $294 - 294 = 0$, we are done."

It is easy to check that $37 \cdot 42 = 1,554$, but many people find the method itself confusing. There is no obvious connection between the given problem and trying to divide 1, 15, and 155 by 42. Further, the reason for subtracting 126 from 155 is not clear, and "bringing down" the 4 seems completely arbitrary.

In recent years a new approach understandable to beginners has become popular. To grasp it, visualize the example above as a problem of sharing 1,554 objects equally among 42 people. Suppose we do not know how to divide, but we can subtract (a necessary prerequisite skill) and multiply by 10. Since $10 \cdot 42 = 420$, which is less than 1,554, we can begin by giving 10 objects to each person. This leaves us $1,554 - 420 = 1,134$ objects left to distribute. If we give 10 more objects to each, we will have $1,134 - 420 = 714$ left to distribute. For a third time we give 10 objects to each, leaving $714 - 420 = 294$ to be distributed. Since now there are fewer than 420 objects left to give out, we cannot give out 10 more to each, but we do have more than 210, which is half of 420, so we can give out 5 more to each. This leaves $294 - 210 = 84$ objects to be distributed. Giving 1 to each person leaves $84 - 42 = 42$, which is just enough to give 1 more to each. Each person has received $10 + 10 + 10 + 5 + 1 + 1 = 37$ objects. This method is understandable but long; it can be summarized this way:

42 people share	1,554	objects
	420	10 to each
objects left	1,134	
	420	10 to each
objects left	714	
	420	10 to each
objects left	294	
	210	5 to each
objects left	84	
	42	1 to each
objects left	42	
	42	1 to each
objects left	0	37 to each total

This will be more concise if we eliminate the words:

$$42 \overline{)1554}$$

420	10
1134	
420	10
714	
420	10
294	
210	5
84	
42	1
42	
42	1
0	37 total

If we realized at the start that at least 30 objects could be given to each of the people and that when 294 remained there were just enough to give 7 more to each, the work would look like this:

$$42 \overline{)1554}$$

1260	30
294	
294	7
0	37

This is virtually the same as Fig. 6.1, which was so confusing. Now do you understand why subtraction is involved? Do you see why in Fig. 6.1 a 4 is "brought down"?

As usually taught by rote, the division algorithm is presented in a form which is handy for some applications like dividing beyond the decimal point (Sec. 10.3). Students should eventually learn this, but they should also understand the division algorithm in terms of repeated subtraction; only then can any streamlined format make sense.

What if we had to distribute 1,559 objects equally among 42 people? We have seen that giving 37 to each uses 1,554 objects, so 5 are left to give out. These 5 objects may be cut into parts and, as we shall see in Chap. 7, the quotient in this case is not an integer, but a fraction. For now, however, we shall simply consider the 5 extra objects as a *remainder*.

Exercises 6.5

1 Find the remainders in each case.
    (a) $1^2 \div 4$, $3^2 \div 4$, $5^2 \div 4$, $7^2 \div 4$, $9^2 \div 4$
    (b) $2^2 \div 4$, $4^2 \div 4$, $6^2 \div 4$, $8^2 \div 4$, $10^2 \div 4$
    (c) Generalize this, and try to explain it.

2   Some odd primes can be expressed as the sum of two squares, but others cannot. For example,
$5 = 1^2 + 2^2$ and $29 = 2^2 + 5^2$, but there is no way to write 7 or 11 as the sum of two squares.
(a) Experimentally, find which odd primes less than 50 can be written as the sum of two
squares and which cannot.
(b) Divide each odd prime less than 50 by 4. How do the remainders relate to part (a)?
(c) Looking over your work in part (a), can you find any case in which you were able to express
an odd prime as a sum of two *even* squares? Were you able to express any odd primes as the
sum of two odd squares? What can you conclude?
(d) Try to explain your conclusion from part (c). *Hint*: Could the sum of two even numbers or
two odd numbers be odd?
*(e) How does Prob. 1 above, together with part (c) of this problem, partially explain the result
of part (b)?

3   The *digital root* of a number is found by adding its digits, then adding the digits of the result,
and repeating the process until only one digit is left. For example the digital root of 678 is 3,
since $6 + 7 + 8 = 21$ and the digits of 21 add to 3.
(a) The digital root of 32 is ____.
If you divide 32 by 9, the remainder is ____.
(b) The digital root of 314 is ____.
If you divide 314 by 9, the remainder is ____.
(c) The digital root of 582 is ____.
If you divide 582 by 9, the remainder is ____.
*(d) Parts (a), (b), and (c) suggest a general pattern. Find it and test it in several more cases. Are
there any exceptions? (Try it on 9 itself, for example.)
*(e) Can you explain the pattern in part (d)? *Hint*:

$$678 = 6 \cdot 100 + 7 \cdot 10 + 8$$
$$= 6(99 + 1) + 7(9 + 1) + 8$$
$$= 6 \cdot 99 + 6 + 7 \cdot 9 + 7 + 8$$
$$= \underline{6 \cdot 99 + 7 \cdot 9} + 6 + 7 + 8$$

Here the underlined number is a multiple of 9. Why? What about the remainder? Can you do
this with other numbers?

*4   Problem 3 is the basis for *casting out 9s*, which is sometimes used to check computations. (For
this we count digital roots of 9 and 0 as the same.) To see how it works, let's check the addition:

		Add digital roots, because we are
3,877	Digital root of 3,887 is 7	checking *addition*
+ 4,827	Digital root of 4,827 is 3	
	$7 + 3 = 10$	Digital root is $1 + 0 = ①$
8,704	Digital root of 8,704 is ①	These are the same, so addition checks.

To check multiplication, proceed the same way, except that instead of adding digital roots you
multiply them.
(a) One or two of these computations are wrong. Use casting out 9s to find which.

48	237	5217
× 21	+ 149	× 2714
1,008	376	14,159,938

(b) Can you use casting out 9s to check subtraction? If so, how?
(c) If you try to use casting out 9s to check division, a special procedure is needed to deal with
remainders. Can you find it?
(d) Is it possible to make a mistake which casting out 9s will not detect? *Hint*: Compare the
digital roots of 8,123 and 8,132.

$2^3 - 2$ is an exact multiple of 3

$2^5 - 2$ is an exact multiple of 5

$2^7 - 2$ is an exact multiple of 7

$2^{11} - 2$ is an exact multiple of 11

(*b*) Verify that

$2^4 - 2$ is not a multiple of 4

$2^6 - 2$ is not a multiple of 6

$2^8 - 2$ is not a multiple of 8

$2^9 - 2$ is not a multiple of 9

$2^{10} - 2$ is not a multiple of 10

(*c*) What general conclusion do parts (*a*) and (*b*) suggest? *Hint*: What distinguishes the exponents in part (*a*) from those in part (*b*)?

(*d*) The ancient Chinese thought (as you probably do) that $2^n - 2$ is a multiple of *n* whenever *n* is prime but never when *n* is composite. In this they were partly wrong. If *n* is prime, $2^n - 2$ is indeed a multiple of *n*, but there are also some composite numbers *n* for which $2^n - 2$ is a multiple of *n*. The smallest such number is 341. What are its prime factors?

6 (*a*) Choose a 3-digit number and copy it next to itself to form a 6-digit number. For example, if your number was 382, you now have 382,382.

(*b*) Divide your 6-digit number by 7. (There will be no remainder.)

(*c*) Divide the quotient from part (*b*) by 11. (Again there will be no remainder.)

(*d*) Divide the quotient from part (*c*) by 13. What do you notice?

(*e*) Can you explain this? *Hint*: What is $7 \cdot 11 \cdot 13$? Try multiplying this by your 3-digit number.

7 646,464 is a multiple of 7, as are 303,030 and 787,878. In fact, every 6-digit number whose digits are in the pattern *ababab* is a multiple of 7. Why? *Hint*: Compute $646,464 \div 64$ and $787,878 \div 78$; then factor.

*8 Fermat (see page 147) observed that:

$2^{2^1} + 1 = 5$

$2^{2^2} + 1 = 2^4 + 1 = 17$

$2^{2^3} + 1 = 2^8 + 1 = 257$

$2^{2^4} + 1 = 2^{16} + 1 = 65,537$

are all primes, and he guessed that every number of form $2^{2^n} + 1$ is prime. But Euler (see page 203) observed that $2^{2^5} + 1 = 2^{32} + 1 = 4,294,967,297$ is a multiple of 641 and so is not prime. Verify Euler's assertion.

CHAPTER 7
FRACTIONS

So far we have dealt only with whole numbers, but now we must widen our scope to include fractions. You may be pleasantly surprised to find that after our rather thorough review of integers, fractions are not as difficult as you expected.

7.1 INTRODUCTION TO FRACTIONS

You probably have an intuitive feeling for fractions, built up from years of using them in measurements of various kinds and from diagrams like those in Fig. 7.1. But what are fractions really? This question is deeper than you might think; a full answer would take us far afield. Here we simply define the fraction $\frac{a}{b}$ to mean $a \div b$. (Note that the division sign looks like an abstract picture of a fraction.) In the fraction a/b, we call a the *numerator* and b the *denominator*. Since division by 0 is not defined (see Sec. 6.4), there are no fractions whose denominators are 0.

Exercises 7.1

1 Express as a fraction:
   (a) $11 \div 3$ (b) $-8 \div 4$ (c) $15 \div (-4)$ (d) $-117 \div (-49)$

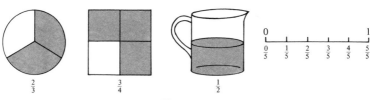

Figure 7.1

2 Express in the form $a \div b$:

(a) $\frac{2}{3}$ (b) $\frac{5}{9}$ (c) $\frac{-14}{21}$ (d) $\frac{-4}{-7}$

3 What fraction of each shape is shaded?

(a) 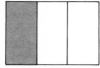   (b)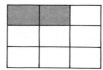

(c)   (d)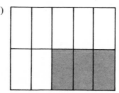

(e)   (f)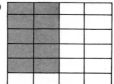

(g)

(h)

4 Label the points of division which are marked but not labeled.

(a)

(b)

5 (a) Here is a line segment 2 units long, cut into 3 equal pieces:

How long is each piece?

(b) This line segment is ____ units long, and it is cut into ____ equal pieces. How long is each?

6 Fill in the blanks, using the fact that a/b means $a \div b$.

(a) $\dfrac{7}{7} =$ _____   (b) $\dfrac{100}{100} =$ _____   (c) $\dfrac{-3}{-3} =$ _____   (d) $\dfrac{}{-1} = 1$

(e) $1 = \dfrac{5}{}$ (f) $\dfrac{}{3} = 1$ (g) generalize

7.2 HOW ARE FRACTIONS MULTIPLIED?

You can find out for yourself how to multiply fractions by working through the following examples. They are based on the geometric view of multiplication; if a rectangle is ℓ units long and w units wide, there are ℓw square units of area in it.

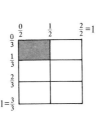
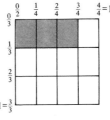

Figure 7.2   **Figure 7.3**

The square in Fig. 7.2 is 1 unit long and 1 unit wide, so it has 1 square unit of area in it. Then how many square units of area are in the shaded rectangle? _____ What are the length and width of the shaded rectangle? _____ , _____ . This shows that $\frac{1}{2} \cdot \frac{1}{3} = \frac{1}{6}$. *Note*: The product is smaller than either factor. That is not unusual and need not bother you.

How many small subrectangles are in the square in Fig. 7.3? _____ The area in each subrectangle is _____ square unit(s). The area in the shaded rectangle is _____ square units. Length and width of the shaded rectangle: _____ , _____ . Multiplication illustrated:

$$\frac{}{3} \cdot \frac{}{4} = \frac{}{12}$$

The area in each subrectangle in Fig. 7.4 is _____ square unit(s). Area in shaded rectangle: _____ square unit(s). Length and width of shaded rectangle: _____ , _____ . Multiplication illustrated: _____ × _____ = _____ .

Some labels in Fig. 7.5 are deliberately left out. Area in each subrectangle: _____ square unit(s). Area in shaded rectangle: _____ square unit(s). Length and width of shaded rectangle: _____ , _____ . Multiplication illustrated: _____ × _____ = _____ . In Fig. 7.6 the unit square is marked with a heavy border. How many small rectangles are in the unit square? _____ . Area in each small rectangle? _____ . Area in the shaded rectangle? _____ . Length and width of shaded rectangle: _____ , _____ . Multiplication illustrated: _____ × _____ = _____ .

Can you make similar diagrams to illustrate other multiplications? Can

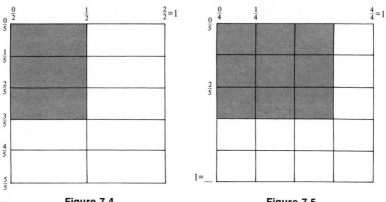

Figure 7.4   **Figure 7.5**

you see how to multiply fractions without drawing diagrams at all? If you need more practice or want to check yourself, the following exercises will help.

Exercises 7.2

1 Make diagrams like Figs. 7.2 to 7.6 to show and compute:

    (a) $\dfrac{1}{5} \cdot \dfrac{1}{3}$ (b) $\dfrac{3}{4} \cdot \dfrac{1}{7}$ (c) $\dfrac{9}{4} \cdot \dfrac{3}{5}$ (d) $\dfrac{6}{5} \cdot \dfrac{7}{5}$

2 Find these products without using diagrams:

    (a) $\dfrac{7}{9} \cdot \dfrac{5}{10}$ (b) $\dfrac{7}{11} \cdot \dfrac{15}{13}$ (c) $\dfrac{100}{203} \cdot \dfrac{47}{51}$ (d) $\dfrac{p}{q} \cdot \dfrac{r}{s}$

7.3 EQUAL FRACTIONS

More than one fraction can represent a given point on a number line, as Fig. 7.7 shows. Fractions which represent the same point on a number line are said to be *equal*. Figure 7.7 shows that $\frac{3}{2}$, $\frac{6}{4}$, and $\frac{12}{8}$ are equal, for example.

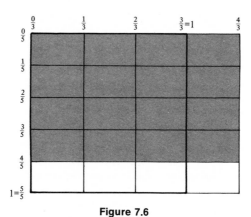

Figure 7.6

Figure 7.7

The fractions which equal 1 are an especially important family, which we shall refer to often. You met some of these in Prob. 6 of Exercises 7.1, and Fig. 7.7 shows others. In general, if n is any number except 0, the fraction n/n equals 1. This is true even if n is itself a fraction.

Multiplying by 1 does not change any number, even if the factor of 1 is expressed as a fraction. Therefore, to find fractions equal to a given fraction, multiply the given fraction by 1 in the form of a fraction. Thus, $\frac{1}{2} \cdot \frac{2}{2} = \frac{2}{4}$, and $\frac{1}{2} \cdot \frac{4}{4} = \frac{4}{8}$, confirming, as Fig. 7.7 shows, that $\frac{1}{2}$, $\frac{2}{4}$, and $\frac{4}{8}$ are all equal.

Exercises 7.3

1 (a) $\dfrac{3}{5} \cdot \dfrac{n}{n} = \dfrac{6}{10}$ $n = $ _____   (b) $\dfrac{3}{5} \cdot \dfrac{m}{m} = \dfrac{-9}{-15}$ $m = $ _____

 (c) $\dfrac{3}{5} \cdot \dfrac{k}{k} = \dfrac{150}{250}$ $k = $ _____

2 (a) $\dfrac{11}{6} \cdot \dfrac{h}{h} = \dfrac{77}{42}$ $h = $ _____   (b) $\dfrac{11}{6} \cdot \dfrac{t}{t} = \dfrac{110}{60}$ $t = $ _____

 (c) $\dfrac{11}{6} \cdot \dfrac{u}{u} = \dfrac{-143}{-78}$ $u = $ _____

3 (a) $\dfrac{5}{-12} \cdot \dfrac{-1}{-1} =$ (b) $\dfrac{13}{-19} \cdot \dfrac{-1}{-1} =$

 (c) $\dfrac{-18}{11} \cdot \dfrac{-1}{-1} =$ (d) generalize

4 Which of these, if any, equal $\dfrac{4}{7}$? (Find which result from multiplying $\dfrac{4}{7}$ by a form of 1.)

 (a) $\dfrac{12}{21}$ (b) $\dfrac{13}{22}$ (c) $\dfrac{-40}{-70}$ (d) $\dfrac{-76}{-133}$

5 Fill in the blanks.

 (a) $\dfrac{2}{9} = \dfrac{}{27} = \dfrac{10}{} = \dfrac{}{90} = \dfrac{2,000}{}$ (b) $\dfrac{-3}{10} = \dfrac{}{20} = \dfrac{-15}{} = \dfrac{27}{}$

7.4 SIMPLIFYING FRACTIONS

Which of these fractions is simplest?

$$\dfrac{3}{12} \qquad \dfrac{36}{144} \qquad \dfrac{251}{1,004} \qquad \dfrac{1}{4} \qquad \dfrac{19}{76}$$

They are all equal, but most people choose $\frac{1}{4}$, as 1 and 4 are small and therefore easy to think of. From this point of view a fraction which is expressed as simply as possible is said to be in lowest terms. More formally, a fraction is in *lowest terms* if numerator and denominator are not both multiples of an integer greater than 1. Is $\frac{6}{21}$ in lowest terms? No, as 6 and 21 are both multiples of 3. To reduce $\frac{6}{21}$ to lowest terms,

$$\frac{6}{21} = \frac{2 \cdot 3}{7 \cdot 3} = \frac{2}{7} \cdot \frac{3}{3} = \frac{2}{7} \cdot 1 = \frac{2}{7}$$

Any fraction may be treated similarly, unless it is in lowest terms to begin with. The idea is to use the highest factor common to the numerator and denominator, together with the fact that multiplying by 1 in any form does not change the value of any number. For example, to simplify $\frac{18}{42}$, we write

$$\frac{18}{42} = \frac{3 \cdot 6}{7 \cdot 6} = \frac{3}{7} \cdot \frac{6}{6} = \frac{3}{7} \cdot 1 = \frac{3}{7}$$

Here it was not very difficult to see that 6 is the highest common factor of 18 and 42, but when large numbers are involved, a more systematic approach is needed. You will find one in Prob. 8 below.

Reducing fractions in advance can sometimes save computation. Suppose, for instance, you must multiply $\frac{53}{108}$ by $\frac{9}{106}$. The tedious computation yields $\frac{477}{11448}$, which is, to put it mildly, clumsy. This can be avoided as follows:

$$\frac{53}{108} \cdot \frac{9}{106} = \frac{53}{12 \cdot 9} \cdot \frac{9}{53 \cdot 2} = \frac{53 \cdot 9}{12 \cdot 2 \cdot 53 \cdot 9}$$

$$= \frac{1}{12 \cdot 2} \cdot \frac{53 \cdot 9}{53 \cdot 9}$$

Since $\dfrac{53 \cdot 9}{53 \cdot 9}$ is a form of 1, we are left with

$$\frac{1}{12 \cdot 2} = \frac{1}{24}$$

This shortcut used to be called canceling, and mathematicians still call it that, but this bit of mathematical slang was abused by people who did not really understand the process, so the term led to confusion and went out of style.

Exercises 7.4

1 Express each of these in lowest terms.

(*a*) $\dfrac{1+3}{5+7}$ (*b*) $\dfrac{1+3+5}{7+9+11}$ (*c*) $\dfrac{1+3+5+7}{9+11+13+15}$ (*d*) generalize

2 Express this product as a single fraction in lowest terms.

$$\frac{1}{2} \cdot \frac{2}{3} \cdot \frac{3}{4} \cdot \frac{4}{5} \cdot \frac{5}{6} \cdot \frac{6}{7} \cdot \frac{7}{8}$$

(Look for shortcuts!)

3 Fill in the blanks.

(a) $\dfrac{1+2+3}{4+5+6} = \dfrac{}{5}$　　(b) $\dfrac{2+3+4}{5+6+7} = \dfrac{}{6}$　　(c) $\dfrac{3+4+5}{6+7+8} = \dfrac{}{7}$

(d) What do you predict for $\dfrac{4+5+5}{7+8+9}$? Check your prediction.

4 Fill in the blanks.

(a) $\dfrac{1-2+3}{4-5+6} = \dfrac{}{5}$　　(b $\dfrac{2-3+4}{5-6+7} = \dfrac{}{6}$　　(c) $\dfrac{3-4+5}{6-7+8} = \dfrac{}{7}$

(d) What do you predict for $\dfrac{4-5+6}{7-8+9}$?

5 Multiply, looking for shortcuts.

(a) $\dfrac{5}{18} \cdot \dfrac{-6}{15}$　　(b) $\dfrac{21}{25} \cdot \dfrac{20}{42}$　　(c) $\dfrac{-37}{50} \cdot \dfrac{10}{74}$

⋆(d) $\dfrac{-35}{39} \cdot \dfrac{91}{75} \cdot \dfrac{5}{7}$

6 Express each product as simply as possible.

(a) $\dfrac{4}{1} \cdot \dfrac{1}{4}$　　(b) $\dfrac{17}{1} \cdot \dfrac{1}{17}$　　(c) $\dfrac{1}{-5} \cdot \dfrac{-5}{1}$　　(d) $\dfrac{3}{5} \cdot \dfrac{5}{3}$　　(e) generalize

7 Fill in the missing numbers.

(a) $\dfrac{3}{5} \cdot \dfrac{}{} = \dfrac{15}{15} = 1$　　　　(b) $\dfrac{100}{1} \cdot \dfrac{}{} = \dfrac{100}{100} = 1$

(c) $\dfrac{-9}{7} \cdot \dfrac{}{} = \dfrac{-63}{-63} = 1$　　(d) $\dfrac{1}{-19} \cdot \dfrac{}{} = \dfrac{-19}{-19} = 1$

(e) generalize

8 Here is a way to find the highest common factor (h c f) of two positive integers. Call the smalle of these x and the other y.

> **1.** Divide y by x and call the remainder r_1.
> If $r_1 = 0$, x is the hcf of x and y. Otherwise:
> **2.** Divide x by r_1 and call the remainder r_2.
> If $r_2 = 0$, r_1 is the hcf of x and y. Otherwise:
> **3.** Divide r_1 by r_2 and call the remainder r_3.
> If $r_3 = 0$, r_2 is the hcf of x and y. Otherwise:
> **4.** Continue this way. The last nonzero remainder you get is the hcf of x and y.

For example to find the hcf of 112 and 148:

> **1.** Divide 148 by 112. The reminder is $148 - 112 = 36$.
> **2.** Divide 112 by 36. The reminder is $112 - 3 \cdot 36 = 112 - 108 = 4$.
> **3.** Divide 36 by 4. The remainder is 0, and so 4 is the hcf of 148 and 112.

Use this method to express in lowest terms:

(a) $\dfrac{150}{168}$　　(b) $\dfrac{393}{429}$　　(c) $\dfrac{792}{838}$

7.5 DIVISION OF FRACTIONS

To divide by a fraction, countless people have been taught to invert the divisor and multiply. For example, to divide $\frac{5}{7}$ by $\frac{2}{3}$, first invert $\frac{2}{3}$ to get $\frac{3}{2}$, then multiply: $\frac{5}{7} \div \frac{2}{3} = \frac{5}{7} \cdot \frac{3}{2} = \frac{15}{14}$.

Is this correct? Like any division, it can be checked by multiplying, and indeed, $\frac{2}{3} \cdot \frac{15}{14} = \frac{5}{7}$. Still, inverting the divisor and multiplying seems more like magic than mathematics. You can see what is really going on if you bear in mind three facts, all of which are already familiar to you:

> **1.** Multiplication or division by 1 does not change the value of any number.
>
> **2.** $\frac{2}{3} \cdot \frac{3}{2} = 1$
>
> **3.** $\dfrac{\frac{3}{2}}{\frac{3}{2}} = 1$

Then

$$\frac{5}{7} \div \frac{2}{3} = \frac{\frac{5}{7}}{\frac{2}{3}} = \frac{\frac{5}{7}}{\frac{2}{3}} \cdot \frac{\frac{3}{2}}{\frac{3}{2}} = \frac{\frac{5}{7} \cdot \frac{3}{2}}{\frac{2}{3} \cdot \frac{3}{2}} = \frac{\frac{5}{7} \cdot \frac{3}{2}}{1} = \frac{5}{7} \cdot \frac{3}{2}$$

This reasoning applies to division by any fraction, provided that it is not equal to 0. The number obtained by inverting a fraction is called its *reciprocal* or *multiplicative inverse*.

Exercises 7.5

1　Fill in the blanks.

(a) $\dfrac{3}{7} \div \dfrac{4}{5} = \dfrac{\frac{3}{7}}{\frac{4}{5}} = \dfrac{\frac{3}{7}}{\frac{4}{5}} \cdot \dfrac{\frac{5}{4}}{\frac{5}{4}} = \dfrac{\overline{\quad}}{\frac{28}{20}} = \dfrac{\overline{\quad}}{1} = \dfrac{\overline{\quad}}{\quad}$

(b) $\dfrac{5}{7} \div \dfrac{4}{9} = \dfrac{\overline{\quad}}{\frac{4}{9}} = \dfrac{\frac{5}{\quad}}{\overline{\quad}} \cdot \dfrac{\overline{\quad}}{\frac{9}{\quad}} = \dfrac{\overline{\quad}}{36} = \dfrac{\overline{\quad}}{\quad}$

(c) $\dfrac{-3}{11} \div \dfrac{7}{15} = \dfrac{\overline{\frac{\overline{11}}{\quad}}}{\frac{7}{\quad}} = \dfrac{\overline{\quad}}{7} \cdot \dfrac{\overline{\quad}}{\overline{\quad}} = \dfrac{\overline{\quad}}{\overline{\quad}} = \dfrac{\overline{\quad}}{\quad}$

(d) $\dfrac{-17}{-4} \div \dfrac{-6}{5} = \dfrac{\overline{\quad}}{\overline{\quad}} = \dfrac{\frac{-17}{\quad}}{\frac{\quad}{5}} \cdot \dfrac{\overline{\quad}}{\overline{\quad}} = \dfrac{\overline{\quad}}{\overline{\quad}} = \dfrac{\overline{\quad}}{\quad}$

2 Divide as indicated and check by multiplying. (If you need to, write your work in full, as in Prob. 1. Do not write more than you need, however.)

(a) $\dfrac{5}{8} \div \dfrac{1}{4}$ (b) $\dfrac{1}{4} \div \dfrac{5}{8}$ (c) $\dfrac{1}{2} \div \dfrac{1}{3}$ (d) $\dfrac{1}{3} \div \dfrac{1}{2}$

(e) $\dfrac{-3}{4} \div \dfrac{4}{5}$ (f) $\dfrac{4}{5} \div \dfrac{-3}{4}$ (g) generalize the pattern from these pairs of examples.

7.6 ADDING FRACTIONS

Fractions are easy to add if they happen to have the same denominator. For example, $\frac{2}{7} + \frac{3}{7}$ should be $\frac{5}{7}$, at least if we believe our intuition and pictures like Fig. 7.8.

The distributive law bears this out. Observe first that $\frac{2}{7}$ can be written $\frac{2}{1} \cdot \frac{1}{7}$ or $2 \cdot \frac{1}{7}$, and $\frac{3}{7} = \frac{3}{1} \cdot \frac{1}{7}$ or $3 \cdot \frac{1}{7}$. Then

$$\frac{2}{7} + \frac{3}{7} = 2 \cdot \frac{1}{7} + 3 \cdot \frac{1}{7} = (2 + 3)\frac{1}{7} = 5 \cdot \frac{1}{7} = \frac{5}{7}$$

This reasoning is entirely general. If a/b and c/b are any two fractions with a given denominator, their sum is

$$\frac{a + c}{b}$$

What if the denominators are not all the same? Then the situation is a bit more complex, as we shall see. Essentially, the idea is to arrange an equivalent addition of fractions with a common denominator. As an example, consider $\frac{5}{6} + \frac{1}{4}$. Since $\frac{5}{6} = \frac{10}{12}$ and $\frac{1}{4} = \frac{3}{12}$, we can rewrite the problem as $\frac{10}{12} + \frac{3}{12}$, which is $\frac{13}{12}$.

How was 12 chosen as common denominator? It is the least common multiple of the original denominators, 6 and 4. Any common multiple of 6 and 4 would do, but using the *least* common multiple avoids unnecessary complications.

The above example is entirely typical, provided all the fractions to be added have positive denominators. If any fractions have negative denominators, we first change them for equal fractions with positive denominators. An example of this is

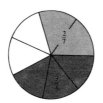

Figure 7.8

$$\frac{3}{4} + \frac{4}{-5} + \frac{1}{6}$$

Here we first rewrite $\dfrac{4}{-5}$ as $\dfrac{4}{-5} \cdot \dfrac{-1}{-1} = \dfrac{-4}{5}$

then add $\dfrac{3}{4} + \dfrac{-4}{5} + \dfrac{1}{6}$

as before, using 60, the least common multiple of 4, 5, and 6, as common denominator:

$$\frac{3}{4} = \frac{45}{60} \qquad \frac{-4}{5} = \frac{-48}{60} \quad \text{and} \quad \frac{1}{6} = \frac{10}{60}$$

Therefore,

$$\frac{3}{4} + \frac{-4}{5} + \frac{1}{6} = \frac{45 + (-48) + 10}{60} = \frac{7}{60}$$

Looking back, it is easy to see why elementary school students have trouble learning to add fractions, for they need to know first:

1.   How to find the least common multiple of several numbers (this in turn demands understanding of primes and factoring)
2.   How to express a given fraction as an equal fraction with a predetermined denominator
3.   How to add fractions with a common denominator

Most students need several years to build the background needed to understand addition of fractions. Teachers should not attempt to teach this or any other topic to children who lack prerequisite knowledge, for the most likely result will be confused students who dislike mathematics and cause discipline problems.

Exercises 7.6

1   Add:

(a) $\dfrac{1}{2} + \dfrac{1}{4}$ (b) $\dfrac{1}{2} + \dfrac{1}{4} + \dfrac{1}{8}$ (c) $\dfrac{1}{2} + \dfrac{1}{4} + \dfrac{1}{8} + \dfrac{1}{16}$

(d) $\dfrac{1}{2} + \dfrac{1}{4} + \dfrac{1}{8} + \dfrac{1}{16} + \dfrac{1}{32}$ (e) generalize

2   Add:

(a) $\dfrac{1}{3} + \dfrac{1}{9}$ (b) $\dfrac{1}{3} + \dfrac{1}{9} + \dfrac{1}{27}$ (c) $\dfrac{1}{3} + \dfrac{1}{9} + \dfrac{1}{27} + \dfrac{1}{81}$ (d) generalize

3 Add:

(a) $\dfrac{1}{2} + \dfrac{1}{3}$ (b) $\dfrac{1}{3} + \dfrac{1}{4}$ (c) $\dfrac{1}{4} + \dfrac{1}{5}$ (d) $\dfrac{1}{5} + \dfrac{1}{6}$ (e) generalize

4 Add:

(a) $\dfrac{1}{1 \cdot 2} + \dfrac{1}{2 \cdot 3}$ (b) $\dfrac{1}{2 \cdot 3} + \dfrac{1}{3 \cdot 4}$ (c) $\dfrac{1}{3 \cdot 4} + \dfrac{1}{4 \cdot 5}$

(d) $\dfrac{1}{4 \cdot 5} + \dfrac{1}{5 \cdot 6}$ (e) generalize

Hint: The pattern is clearest if you express each answer with 2 as numerator.

5 Add:

(a) $\dfrac{-3}{4} + \dfrac{3}{4}$ (b) $\dfrac{11}{5} + \dfrac{-11}{5}$ (c) $\dfrac{97}{83} + \dfrac{-97}{83}$

(d) In general, what is the opposite of the fraction a/b?

6 As with any number, a fraction can be subtracted by adding its opposite. With that and Prob. 5 in mind, subtract:

(a) $\dfrac{7}{8} - \dfrac{2}{3}$ (b) $\dfrac{4}{5} - \dfrac{9}{10}$ (c) $\dfrac{3}{11} - \dfrac{-5}{7}$ (d) $\dfrac{4}{7} - \dfrac{-2}{3}$

7 Measurements are often expressed as *mixed numbers* (part integer, part fraction), such as $99\frac{44}{100}$, which means $99 + \frac{44}{100}$. Any fraction greater than 1 can be expressed this way by carrying out the division; for example,

$$\frac{17}{3} = 5\frac{2}{3}$$

Express as mixed numbers:

(a) $\dfrac{17}{12}$ (b) $\dfrac{22}{7}$ (c) $\dfrac{50}{3}$

8 Express as fractions:

(a) $1\frac{2}{3}$ *Hint*: $1 = \dfrac{3}{3}$ (b) $6\frac{7}{8}$ *Hint*: $6 = \dfrac{?}{8}$

9 A room is rectangular, $12\frac{1}{2}$ feet long and $10\frac{2}{3}$ feet wide. For wall-to-wall carpeting, how many square feet of carpet will be needed?

10 Could fractions be added on an adder like that in Sec. 2.4? What would some of the practical difficulties be in trying to use this idea?

11 Add and express each sum in lowest terms:

(a) $\dfrac{1}{3} + \dfrac{1}{3 \cdot 2}$ (b) $\dfrac{1}{4} + \dfrac{1}{4 \cdot 3}$ (c) $\dfrac{1}{5} + \dfrac{1}{5 \cdot 4}$

(d) $\dfrac{1}{6} + \dfrac{1}{6 \cdot 5}$ (e) $\dfrac{1}{7} + \dfrac{1}{7 \cdot 6}$

(f) From the pattern in parts (a) through (e), predict the sums

$$\frac{1}{20} + \frac{1}{20 \cdot 19} \quad \text{and} \quad \frac{1}{50} + \frac{1}{50 \cdot 49}$$

Then check your predictions.

12 Express each of these products in lowest terms and find a pattern to your answers.

(a) $1 - \frac{1}{2}$ (b) $\left(1 - \frac{1}{2}\right)\left(1 - \frac{1}{3}\right)$ (c) $\left(1 - \frac{1}{2}\right)\left(1 - \frac{1}{3}\right)\left(1 - \frac{1}{4}\right)$

(d) $\left(1 - \frac{1}{2}\right)\left(1 - \frac{1}{3}\right)\left(1 - \frac{1}{4}\right)\left(1 - \frac{1}{5}\right)$ (e) generalize

13 Compute these products:

(a) $1 - \frac{1}{2^2}$ (b) $\left(1 - \frac{1}{2^2}\right)\left(1 - \frac{1}{3^2}\right)$

(c) $\left(1 - \frac{1}{2^2}\right)\left(1 - \frac{1}{3^2}\right)\left(1 - \frac{1}{4^2}\right)$

(d) $\left(1 - \frac{1}{2^2}\right)\left(1 - \frac{1}{3^2}\right)\left(1 - \frac{1}{4^2}\right)\left(1 - \frac{1}{5^2}\right)$

(e) generalize

14 Add:

(a) $\frac{1}{1} + \frac{1}{2} + \frac{1}{6}$ (b) $\frac{1}{1} + \frac{1}{5} + \frac{1}{20}$ (c) $\frac{1}{2} + \frac{1}{3} + \frac{1}{6}$ (d) $\frac{1}{2} + \frac{1}{4} + \frac{1}{12}$

The Polish mathematician Sierpinski has conjectured that any fraction 5/n, where n is at least 3, can be expressed as a sum of just three fractions no two of which are the same and all of whose numerators are 1. Parts (a) to (d) illustrate this in four cases. This conjecture has not been proved, but it is known that there are no exceptions with n less than 1,057,438,801.

15 The first two "perfect" numbers (defined in Sec. 5.3) are 6 and 28. Their factors, respectively, are 1, 2, 3, 6 and 1, 2, 4, 7, 14, 28.

(a) $\frac{1}{1} + \frac{1}{2} + \frac{1}{3} + \frac{1}{6} =$

(b) $\frac{1}{1} + \frac{1}{2} + \frac{1}{4} + \frac{1}{7} + \frac{1}{14} + \frac{1}{28} =$

(c) The next perfect number is 496. Its factors are 1, 2, 4, 8, 16, 31, 62, 124, and 496. On the basis of parts (a) and (b), what do you predict for the sum of the reciprocals of these factors:

$\frac{1}{1} + \frac{1}{2} + \frac{1}{4} + \frac{1}{8} + \frac{1}{16} + \frac{1}{31} + \frac{1}{62} + \frac{1}{124} + \frac{1}{248} + \frac{1}{496} =$

(d) Check your prediction from part (c). Hint: The last five fractions may be written $\frac{1}{31}\left(\frac{1}{1} + \frac{1}{2} + \frac{1}{4} + \frac{1}{8} + \frac{1}{16}\right)$.

CHAPTER 8
GRAPHS AND RULES

Two very important ideas are introduced in this chapter. One is at the heart of our ideas of pattern and relationship, and the other makes it possible to visualize such relationships. These ideas are fascinating in their own right, and they raise still more intriguing questions.

8.1 POINTS AND COORDINATES

Numbers are often used to specify locations. The usual way to do this is by reference to a pair of perpendicular number lines called *axes* (singular, *axis*), which cross at their 0 points, as in Fig. 8.1. The horizontal and vertical axes are usually called the x and y axes, respectively. With reference to given axes, each point is specified by an ordered pair of numbers called *coordinates*, usually written in parentheses and separated by a comma. By convention, the first coordinate refers to the x axis and the second to the y axis. This is as arbitrary as any convention, such as driving on the right-hand side of the road. It could as well be the other way around, but there would be chaos if no convention were used.

Given the coordinates of a point, how can we plot, that is, locate, it? Find the first coordinate on the x axis and from there move vertically to the level indicated by the second coordinate on the y axis. For example, to plot $(-2,3)$, first find -2 on the x axis, then move up to the level of 3 on the vertical axis as shown in Fig. 8.1. A point can be plotted even if its coordinates are not integers, as $\left(3\frac{1}{2},2\frac{1}{2}\right)$ shown in Fig. 8.1.

Exercises 8.1

1 The coordinate axes cut the plane into four regions, labeled I to IV in Fig. 8.2.
  (*a*) If both coordinates of a point are positive, it lies in region ____.

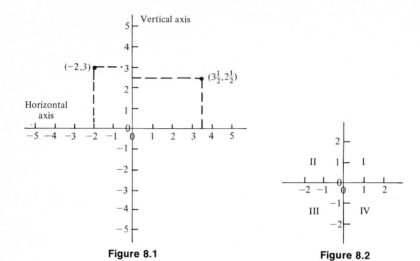

Figure 8.1   Figure 8.2

(*b*) If both coordinates are negative, it lies in region ____ .

(*c*) If the *x* coordinate is negative but the *y* coordinate is positive, the point is in region ____ .

(*d*) If the *x* coordinate is positive but the *y* coordinate is negative, the point is in region ____ .

(*e*) Where is a point whose *x* coordinate is 0?

(*f*) Where is a point whose *y* coordinate is 0?

2   Draw and number coordinate axes, and then plot the following points. (Squared paper, though helpful, is not necessary.)

$(-2,0)$, $(-2,1)$, $(-2,2)$, $(-2,-2)$, $(-2,-1)$, $(-1,0)$, $(0,0)$, $(1,0)$, $(1,1)$, $(1,2)$, $(1,-1)$, $(1,-2)$, $(3,2)$, $(3,1)$, $(3,0)$, $(3,-1)$, $(3,-2)$.

Look for a message in the result.

3   Make up a short message and encode it by reversing the procedure of Prob. 2.

4   Plot these points in the order given, connecting each to the previous by a straight line.

$\left(\frac{1}{2},-3\right)$, $\left(\frac{1}{2},-\frac{1}{2}\right)$, $\left(5\frac{1}{2},-2\right)$, $\left(1,1\frac{1}{2}\right)$, $(4,0)$, $\left(1,2\frac{1}{2}\right)$, $(3,2)$, $(1,4)$, $\left(2,3\frac{1}{2}\right)$, $(0,6)$, $\left(-2,3\frac{1}{2}\right)$, $(-1,4)$, $(-3,2)$, $\left(-1,2\frac{1}{2}\right)$, $(-4,0)$, $\left(-1,1\frac{1}{2}\right)$, $\left(-5\frac{1}{2},-2\right)$, $\left(-\frac{1}{2},-\frac{1}{2}\right)$, $\left(-\frac{1}{2},-3\right)$ $\left(\frac{1}{2},-3\right)$.

8.2 RULES

Rules are best introduced by examples.

Rule A   If this rule is applied to 3 it yields 6, and if it is applied to 7 it yields 14. Can you guess this rule? Can you apply it to 12?

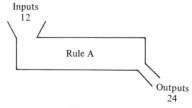

Figure 8.3

A compact way to arrange the information is a table like this:

Applying rule to	3	7	12	4
Yields	6	14	?	8

The table also tells us that applying the rule to 4 yields 8. Does that fit your pattern?

By now you may have guessed that this rule multiplies whatever number it is applied to by 2. You will soon see examples of more complicated rules.

If we think of a rule as a process, we can visualize it as being performed by a machine like the one in Fig. 8.3, which applies the rule to whatever number is put in at the top and sends the output out the bottom.

What if two rules are hooked together so that the outputs of one are used as inputs of the second? For example, suppose the outputs from rule A are used as inputs for rule B:

Rule B Add 7 to whatever number is given.

We can picture the combined operation as in Fig. 8.4. (Can you fill in the question marks?)

A natural shorthand for this rule is $2n + 7$, which means "twice the number plus 7." The expression $2n + 7$ is not itself a number; it is a descriptive name for a rule. But if n is assigned a specific value such as 9, then $2n + 7$ becomes $2 \cdot 9 + 7$ or 25, which is the number the rule associates with 9. From that point of view the symbolism $2n + 7$ is convenient. Here the letter n has no particular significance; any other letter, even from a different alphabet, would do as well. Thus $2n + 7$, $2N + 7$, $2A + 7$, $2\beta + 7$, and $2\, ☺ + 7$ all denote the same rule.

The possibilities for rules are unlimited. One rule might triple whatever number is given, then subtract 17. (It might be written $3N - 17$.) Another might square whatever number is given, then add 4. (It could be written $X^2 + 4$.)

Guessing rules makes a good game. Someone thinks of a rule and others try to guess it. As clues, they ask him to apply the rule to numbers they name. When someone thinks he can guess the rule, he is tested by being asked to

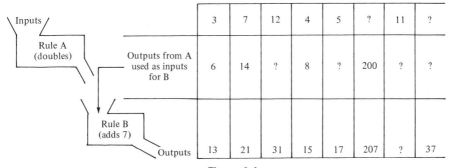

Figure 8.4

apply it to a number. His answer gives another clue to those still groping. The game continues until everyone has guessed the rule.

This game would be worthwhile if it did no more than provide an enjoyable source of arithmetic practice, but in fact it does a great deal more. It is an excellent introduction to the scientific method of inquiry, which George Polya sums up in three syllables as "guess and test." It is a long way from the simple rules discussed here to great scientific achievements such as the discovery that the planets travel, at least approximately, in ellipses about the sun or that the genetic basis of heredity is contained in pairs of spirals intertwined with each other. Remarkable though these are, they are examples of how, with great persistence and ingenuity, scientists have managed to uncover nature's rules.

As it is used here the word "rule" denotes the heart of what mathematicians call a *function*. If P and Q are collections of objects, a *function* from P to Q is a correspondence which associates an object in Q with each object in P. Naturally, to specify a function one must identify both the collections P and Q, as well as the rule of correspondence. In our informal treatment we shall not need to identify P and Q clearly (they will invariably be sets of numbers). Instead we shall concentrate on the rules of correspondence themselves or, as we call them, rules.

Exercises 8.2

1 These rules are given in algebraic shorthand. Describe them in words:

(a) $n + 2$ (b) $3n - 7$ (c) $N^2 + 1$ (d) $(N + 1)^2$ (e) $\dfrac{x + 6}{x - 3}$

(f) $(y - 1)(y + 2)$

2 Guess these rules and fill in the tables (We shall use these rules in later problems).

(a) **Rule C**

Inputs	1	5	13	20		−2
Outputs	7	11	19		6	

Shorthand for rule C _____ .

(b) **Rule D**

Inputs	10	27	0	−2	18		
Outputs	5	$13\frac{1}{2}$	0	−1		−20	−7

Shorthand for rule D _____ .

(c) **Rule E**

Inputs	0	11	100	2	$3\frac{1}{2}$		
Outputs	−7	4	93			−11	−50

Shorthand for rule E _____ .

3 Rule A (from text) followed by rule C (from Prob. 2).

Inputs for A	3	0	5	2			−3
Outputs from A (used as inputs for C)	6	0		20	100		
Outputs from C	12	6	16	10	26	44	−50

(*a*) Fill in the blanks.
(*b*) Write a shorthand name for this compound rule.

4 Rule C (from Prob. 2) followed by rule A (from text).

Inputs for C	3	0	5	12		
Outputs from C (used as inputs for A)	9	6	0			−3
Outputs from A			22		56	−10

(*a*) Fill in the blanks.
(*b*) Write a shorthand name for this compound rule.

5 (*a*) Compare the compound rules in Probs. 3 and 4. Do these rules produce the same results for all inputs?
(*b*) Is there any input for which these rules yield the same output?
(*c*) In general, if compound rules are made up, as in Probs. 3 and 4, by following one rule with another, does it matter which one is applied first?

6 If rule A from the text is followed by rule D from Prob. 2, something interesting happens.

Inputs for rule A	5	8	0	12	$\frac{1}{2}$	4
Outputs from A used as inputs for D	10					
Outputs from D	5					

(*a*) Fill in the blanks.
(*b*) Write a shorthand name for this compound rule.

7 Generalizing Prob. 6, suppose R and S are two rules and we know only the effect of the compound rule R followed by S as shown in this table.

Inputs for R	5	7	−2	72	$\frac{1}{3}$	−19	10	1
?								
Outputs from S	5	7	−2	72	$\frac{1}{3}$	−19	10	1

(*a*) If rule R adds 10, then rule S _____.
(*b*) If rule R adds −91, then rule S _____.
(*c*) If rule R multiplies by 7, then rule S _____.
(*d*) If rule R divides by 3, then rule S _ _____.
(*e*) If rule S subtracts 4, then rule R ___ ___.
(*f*) If rule S divides by −$\frac{1}{2}$, then rule R _ _____.

8 This extends Prob. 7.
(*a*) If rule R adds 3 then multiplies by 2, ru e S _____. *Hint*: To undo a compound process,

you must usually undo the steps in reverse order. For example, to undo the process of putting a sock and shoe on your foot, you must first take off the shoe then take off the sock.

(*b*) If rule R multiplies by 2 then subtracts 4, rule S _____.

(*c*) If rule R adds 7 then divides by 2, rule S _____.

*9 The rule $\dfrac{x+1}{2x-1}$ has a curious property. If this rule is applied to 4, it yields

$$\frac{4+1}{2\cdot 4 - 1} = \frac{5}{7}$$

Now if the rule is applied to $\frac{5}{7}$ it yields

$$\frac{\frac{5}{7}+1}{2\cdot\frac{5}{7}-1} = \frac{\frac{5}{7}+\frac{7}{7}}{\frac{10}{7}-\frac{7}{7}} = \frac{\frac{12}{7}}{\frac{3}{7}} = \frac{12}{7}\cdot\frac{7}{3} = 4$$

Similarly, if this rule is applied to 1 it yields 2.

(*a*) What do you predict this rule will yield if applied to 2?

(*b*) Apply the rule to 2 and test your prediction.

(*c*) Do you see what is curious about this rule?

(*d*) Experiment with this rule, applying it to other numbers to see whether the curious behavior continues.

(*e*) How could this problem be generalized? (What questions would you ask next?)

10 If a dense object like a stone is dropped, it falls about 16 feet in the first second. If it is allowed to fall for 2 seconds, it falls about 64 feet in that time, and if it falls for 3 seconds, it drops a total of about 144 feet. In a 4-second fall the object drops about 256 feet. Can you guess the rule here? How far would a stone fall if it fell free for 10 seconds? (Galileo[1] found this rule around 1590.)

11 This problem is due to John Marks.

(*a*) Lines from one corner of a triangle to the opposite side:

 1 line   2 lines   3 lines   4 lines

1 line cuts the triangle into ____ pieces

2 lines cut the triangle into ____ pieces

3 lines cut the triangle into ____ pieces

4 lines cut the triangle into ____ pieces

Guess the rule: *n* lines, all from the same corner, cut the triangle into ____ pieces.

(*b*) What if lines are drawn from two corners?

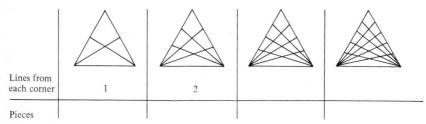

Lines from each corner	1	2		
Pieces				

[1]Galileo Galilei (1564–1642) turned from the study of medicine to science and mathematics, where he quickly won fame. He was a brilliant and popular teacher, and his discoveries, especially those about the pendulum, falling objects, and four moons of Jupiter, show a blend of theory and experiment far ahead of his time. Galileo died blind and discredited, having been forced by the Inquisition to recant because the Church found his discoveries heretical.

Guess the rule: n lines from each of two corners cut the triangle into _____ pieces.

(c) On the basis of the rules from parts (a) and (b), predict what rule would be observed if n lines were drawn from all three corners of the triangle.

(d) Count the pieces in these triangles to check your prediction.

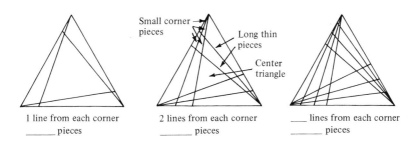

1 line from each corner
_____ pieces

Small corner pieces
Long thin pieces
Center triangle

2 lines from each corner
_____ pieces

_____ lines from each corner
_____ pieces

(e) Do the results in part (d) fit your prediction from part (c)? If not, fill in this table, referring to the pictures.

	Number of lines from each corner			
	0	**1**	**2**	**3**
Number of small pieces at corners	0	3		
Number of long thin pieces	0		6	
Number of center triangles	1			

(f) Guess the rules from the table: If n lines are drawn from each corner:

_____ center triangles will be formed.
_____ long thin pieces will be formed.
_____ small corner pieces will be formed.

Total: _____ pieces will be formed in all.

(g) Test your work from part (f) by drawing 4 lines from each corner (use a large, carefully made sketch) of a triangle and counting the various types of pieces.

(h) For another way to see the rule for the total number of pieces, compare with the second line of part (b) in Prob. 4, Exercises 6.2.

12 Suppose we connect points on a circle by line segments:

(a) With 2 points, as in Fig. 8.5, we can cut the circle into _____ parts.
(b) With 3 points, as in Fig. 8.6, we can cut the circle into _____ parts.
(c) With 4 points, as in Fig. 8.7, we can cut the circle into _____ parts.
(d) Guess the rule. How many parts do you predict could be made by connecting 5 points?
_____ Draw it and see.

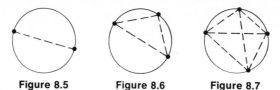

| Figure 8.5 | Figure 8.6 | Figure 8.7 |

(*e*) How many parts do you predict can be made by connecting 6 points? Try it and see. What do you conclude? This shows the value of Polya's warning to test your guesses.

8.3 THE GRAPH OF A RULE

Here is a way to visualize rules by plotting points. To see how it works consider Fig. 8.8, which shows some inputs and outputs for the rule $2n + 7$. With each input number the rule associates an output number, and the resulting ordered pair may be taken as the coordinates of a point. The inputs and outputs in Fig. 8.8 yield the points (0,7), (1,9), (2,11), (3,13), and (4,15). Draw coordinate axes and plot these. What do you notice? Are the points scattered at random? Clearly there is something to investigate here. What questions occur to you? Can you organize them and go about answering them on your own?

This kind of open-ended investigation of a new topic is not easy. At the start one does not know what questions to ask, and it is not clear which observations will have long-range value. It is a little like exploring a new continent; in time names will be given to new discoveries and the relationships between the landmarks will be clarified. Looking back on the mathematical territory now known, you may forget that those who explored it groped, guessed, and followed many blind alleys; today's books contain only that part of their work which succeeded. When you have gone as far as you can on your own (and do not be discouraged if that is not very far), return to this chapter.

Inputs n	Outputs $2n + 7$
0 ⟶	7
1 ⟶	9
2 ⟶	11
3 ⟶	13
4 ⟶	15

Figure 8.8

When you plotted the five points for the rule $2n + 7$, you probably noticed that they all lie on a straight line. Is this coincidence? If the rule $2n + 7$ is applied to other numbers, including negative numbers and fractions, do the resulting points lie on this line? Try a few for yourself, and you will see that indeed they do. In this sense the line on which these points lie is a picture or *graph* of the rule $2n + 7$ (see Fig. 8.9). Is the graph of every rule a straight line? How is the nature of a rule reflected in its graph? The following exercises give some experience that helps deal with such questions.

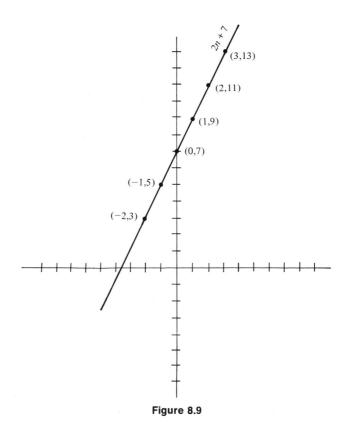

Figure 8.9

Exercises 8.3

1 (*a*) Fill in this table for the rule $3n - 7$.

Inputs n	6	5	3	1	0	-1	-2
Outputs $3n - 7$							

(*b*) Give the coordinates of seven points on the graph of $3n - 7$.
(*c*) Graph $3n - 7$.

2 (*a*) Fill in this table for the rule $2n + \frac{1}{2}$.

Inputs n	-3	-1	$-\frac{1}{4}$	0	$\frac{1}{2}$	1	2
Outputs $2n + \frac{1}{2}$							

(*b*) Graph $2n + \frac{1}{2}$. (You may use the same coordinate axes as in Prob. 1, but graph the new rule distinctively, perhaps with another color.)

3 (*a*) Fill in this table for the rule n^2.

Inputs n	-3	-2	-1	0	1	2	3
Outputs n^2							

(*b*) Give the coordinates of seven points on the graph of n^2.
(*c*) Plot the seven points found in part (*b*).
(*d*) To gain further information about this graph, complete this additional table.

Inputs n	$-2\frac{1}{2}$	$-1\frac{1}{2}$	$-\frac{1}{2}$	$\frac{1}{2}$	$1\frac{1}{2}$	$2\frac{1}{2}$
Outputs n^2						

(e) Plot the resulting additional points on the graph you began in part (c). The graceful curve you are finding is a *parabola*. It is common in mathematics and physics. If air resistance is negligible, as for a dense stone, a thrown object follows a parabola.

4   Graph these rules making up your own inputs and outputs. For efficiency, use several colors to graph several rules on each pair of coordinate axes.

(a) $\frac{1}{2}n + 5$ (b) $-2n + 1$ (c) n^3 (d) $10 - n$

*8.4 LINEAR RULES

A rule is called *linear* if its graph is a straight line. You have already met both nonlinear and linear rules. Examples of linear rules are $2n + 7$, $3n - 7$, and $2n + \frac{1}{2}$; n^2 and n^3 are examples of nonlinear rules. Aside from the fact that their graphs are straight lines, what have these linear rules in common? They all fit the general format $An + B$. For example, if $A = 3$ and $B = -7$, then $An + B$ is $3n + -7$ or $3n - 7$. (What values of A and B make $10 - n$ fit the format $An + B$?) Do all linear rules fit this general pattern? How do the numbers A and B affect the graph of a rule? An experiment will answer these questions.

Using one set of coordinate axes, graph the rules $2n + 3$, $1n + 3$, $\frac{1}{2}n + 3$, $0n + 3$, and $-2n + 3$. The fact that these examples of $An + B$ all have $B = 3$ should correspond to some common property of their graphs, while the different values of A should be related to whatever distinguishes the graphs from each other. Make up your own inputs and outputs, keeping the numbers as easy as possible to avoid unnecessary complications and chances for errors.

The fact that each of these rules fits the general format $An + B$ is reflected in their straight-line graphs. What common property of these lines reflects the fact that in each case $B = 3$? What corresponding property is shared by rules of form $An + B$ with $B = 4$? Here nature has a rule for you to guess. To go even further, you might try to guess the role of A in such rules, though it is more subtle than that of B. Do not read on until you have tried the experiment and pondered these questions on your own.

When you graphed the six rules, you probably noticed that all the graphs pass through the point (0,3). In fact, the graph of any rule $An + B$ passes through (0,B), for if the input value of n is 0, the output is $A \cdot 0 + B = B$. This, then, is the role of B in determining which line goes with which rule. B is sometimes called the y *intercept* because it marks the point where the y axis meets or intercepts the graph of the rule.

There is more to be learned from our experiment. The six rules graphed differ only in their values of A. Geometrically their graphs, six lines through a common point, differ only in direction, so that it seems the value of A in a rule

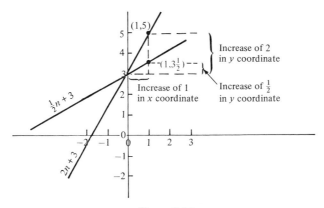

Figure 8.10

$An + B$ determines the direction of the line. Figure 8.10 may help you see how this works. It compares moving from (0,3) to (1,5) along the graph of $2n + 3$ with moving from (0,3) to $\left(1,3\frac{1}{2}\right)$ along the graph of $\frac{1}{2}n + 3$. In both cases the x coordinate is increased by 1, from 0 to 1. But the corresponding change in the y coordinate is an increase of 2 (from 3 to 5) for the rule $2n + 3$ and an increase of only $\frac{1}{2}$ for the rule $\frac{1}{2}n + 3$. In general, suppose we have two points on the graph of $An + B$ and moving from one of these to the other involves an increase of 1 in the x coordinate. Then it also involves an increase of A in the y coordinate. The number A is called the *slope* of the line $An + B$. If A is a large positive number, the graph of $An + B$ points steeply to the upper right; if A is near 0, the graph is nearly horizontal; and if A is negative, the line slopes downward toward the right. Figure 8.11 shows a few examples. With practice it is possible to become so familiar with this "rule of rules" that linear rules can be graphed almost effortlessly. Our more modest goal here, however, is to point out one of the interesting paths for exploration that graphs and rules open up. Pursuing this and related ideas led to the development of large new parts of mathematics in the past four centuries.

8.5 SOME BACKGROUND

Our work has depended heavily on the geometric view of numbers as points on a number line. Relying on that, we have used ordered pairs of numbers to represent points in a plane, and this in turn makes it possible to graph functions or, in our informal treatment, rules. Where does that lead? On the one hand, it has led to the use of algebra as a tool for solving problems in geometry, and on the other hand it has made it possible to use geometry to deal with algebraic questions. The resulting hybrid field, known as *analytic geometry*, has continued to develop in new directions. For example, just as ordered pairs may represent points in a plane, ordered triples of numbers may represent points in

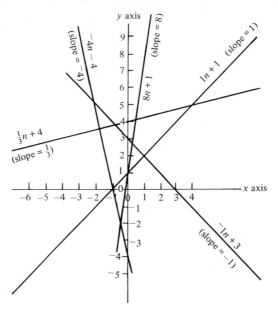

Figure 8.11

three-dimensional space, and this leads to the analytic geometry of three dimensions. The natural extension to ordered quadruples, quintuples, etc., as points in higher-dimensional spaces may sound silly or only academic, since we cannot visualize more than three dimensions, but this study, known as *linear algebra*, is one of the most practical and widely used branches of mathematics.

It was Descartes[2] who first glimpsed these possibilities. True, he probably did not forsee the specific details, but in one sense his original idea was even bigger. In an early unfinished work, "Rules for the Direction of the Mind," he says that the key to all problem solving is to express the problem quantitatively and then apply algebra to work with the resulting equations. Over the next few centuries this approach proved spectacularly successful in the physical sciences, but other areas of inquiry, especially those closest to human life, have been so complex that Descartes' idea seems absurd to many people. Now, however, recent developments in mathematics, together with computers, have made new areas amenable to quantitative study, for example, large parts of economics and psychology. Since these may be just the beginnings, it is too early to say to what extent Descartes' idea was naïve.

[2]René Descartes (1596–1650) was born near Tours, France, and attended the Jesuit school where Mersenne (see page 47) taught. After years soldiering and traveling through Europe, he settled in Holland, where he spent 20 years investigating mathematics, philosophy, and science. His "Discourse on the Method of Rightly Conducting the Reasoning and Seeking Truth in the Sciences," published in 1637, contained the beginnings of plane analytic geometry in an appendix.

CHAPTER 9
EQUATIONS AND WORD PROBLEMS

Equations are one of the main tools for solving problems. We have used equations before, but now we consider them more carefully. First we discuss what equations are and a bit about solving them, and then we see how to use them to solve problems expressed in words.

9.1 INTRODUCTION TO EQUATIONS

An *equation* is a statement that two mathematical objects are the same or, as we say, equal. The verb of such a statement is represented by the sign $=$, which is read "equals" in English. Its meaning is even clearer in German, where it is read "is." (In English we read $3 + 5 = 8$ as "three plus five equals eight," but the German translation of this amounts to "three plus five *is* eight.") Equations may be used to state equality of any mathematical objects, such as numbers, sets, or functions, but for now we shall confine our attention to equations of numbers.

Like any statement, an equation may be true or false. $3(2 + 5) = 21$ is true, since both sides represent the same number, although written differently, but $3 = 4 + 5$ is false, as the two sides are different numbers. An equation like $x + 4 = 10$ cannot be so quickly classified as true or false since that depends on what number x represents. If x is 6, the equation is true, but it is false if x is any other number. This is called a *conditional* equation, since it is true only on condition that x be 6.

A number which satisfies (makes true) a conditional equation is called a *solution* of the equation. $x + 4 = 10$ has only one solution, but some conditional equations have more than one solution, and others have none at all. For example, since $w = w$ is satisfied by any number whatever, it has infinitely many solutions, while $u = u + 1$ has none, since no number is 1 more than itself. These are the extremes; $w = w$ expresses a condition so weak as to be virtually non-existent, while $u = u + 1$ expresses a condition so stringent that no number satisfies it. The most interesting examples lie between these extremes, having some solutions but not many. The exercises below include examples of equations with two and three solutions.

To tell whether or not a given number is a solution of an equation one must simply try it out. The more difficult question of how to find solutions in the first place will be discussed later.

Exercises 9.1

1　(*a*) Is 5 a solution of $3u + 4 = 4u - 1$?
　　(*b*) Is 7 a solution of $x^2 - 5x = 3x - 7$?
　　(*c*) Is 9 a solution of $17a - 4 = 20a - 15$?
　　(*d*) Is -2 a solution of $15k + 3 = 8k - 11$?
　　(*e*) Is $\frac{2}{3}$ a solution of $3p + 4 = 9p$?
　　(*f*) Is $-\frac{1}{4}$ a solution of $\frac{5}{12}n + 13 = -\frac{2}{3}n + 7$?
2　$z^2 + 20 = 9z$ is an example of a conditional equation with two solutions. Both are integers between 1 and 10; find them by trial and error. (You do not need to know how to solve the equation to do this problem.)
3　$y(y - 1)(y - 2) = 0$ is an example of an equation with three solutions. These solutions are all integers between -4 and 4. Find them by trial and error.

9.2 SOLVING EQUATIONS

To *solve* a conditional equation is to find all its solutions. Some equations are very hard to solve, and the development of methods for solving equations has long been important in mathematics. Our goal here is not to make you an expert at solving equations (that would go far beyond this book) but to introduce some general principles which are interesting in their own right and which will enable you to solve simple equations.

To illustrate, we solve $3b + 11 = 5b + 3$ by means of an imaginary experiment. Imagine stones, some sealed in bags and some loose, and suppose that all bags contain the same number of stones. Then $3b + 11 = 5b + 3$ means that the number of stones in three bags, together with eleven loose stones, is the same as the number of stones in five bags, together with three loose stones. Pictorially,

Solving the equation amounts to finding out how many stones there are to a bag without looking inside any bags. Suppose we remove three stones from each side. Now each side has less than before, but it is still true that there are as many stones in and out of bags on the left side as on the right, so that

$$ \text{ᗘᗘᗘ 8888} = \text{ᗘᗘᗘᗘᗘ} $$

This corresponds to the equation $3b + 8 = 5b$. We can simplify this further by removing three bags from each side. Since we do not know how many stones are in each bag, we cannot say how many stones we are removing at this stage, but that does not matter. The crucial point is that there must still be the same number of stones on both sides, so

$$ \text{8888} = \text{ᗘᗘ} $$

or, algebraically, $8 = 2b$. Now we are trying to get an equation which explicitly states what b is, and we have one which states what $2b$ is. To get b from $2b$ we must undo the multiplication by 2 by dividing by 2. Doing this to both sides yields

$$ \text{OOOO} = \text{ᗘ} $$

or $4 = b$. It is easy to check that 4 does indeed satisfy the original equation.

The process of solving $3b + 11 = 5b + 3$ can be summed up as in Fig. 9.1.

The general strategy for solving an equation is to perform a succession of simplifications until an equation is reached which explicitly states a solution. The only rule for constructing each equation from the one before is that whatever is done to one side must be done to the other.

There remains the tactical question of how to decide what operation to perform on both sides to simplify the equation. For this there is no simple rule; often there are several ways to solve the same equation. Skill comes with practice; the following exercises will help.

Exercises 9.2

1   Make a picture with bags and stones which represents:
     (a) $2s + 5$ (b) $4a + 7$ (c) $5g + 1$
2   Represent these equations pictorially with bags and stones:
     (a) $3y + 7 = 7y + 2$ (b) $z + 11 = 5z + 1$

Figure 9.1

	Picture			Algebra
Original equation	ᗘᗘᗘ 8888888	=	ᗘᗘᗘᗘᗘ OOO	$3b + 11 = 5b + 3$
Remove OOO	−OOO		−OOO	$- 3 \qquad - 3$
Result	ᗘᗘᗘ OOOO OOOO	=	ᗘᗘᗘᗘᗘ	$3b + 8 = 5b$
Remove ᗘᗘᗘ	−ᗘᗘᗘ		− ᗘᗘᗘ	$- 3b \qquad - 3b$
Result	8888	=	ᗘᗘ	$8 = 2b$
Cut each side in half	8888	=	ᗘᗘ	$\dfrac{8}{2} = \dfrac{2b}{2}$
Result	88	=	ᗘ	$4 = b$

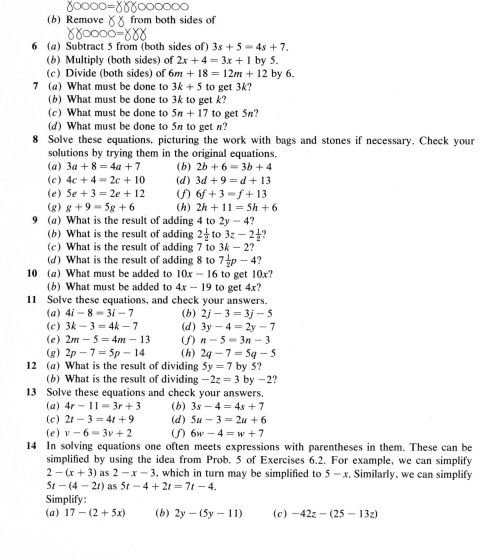

3 If x represents the number of stones in a bag, state each of these pictorial equations algebraically.

(a) ⚬⚬⚬○○=⚬⚬⚬⚬⚬

(b) ⚬⚬⚬⚬⚬⚬○○○○○ = ⚬⚬ ○○○○○○○○○

4 What is the result of:

(a) Removing ○○ from ⚬⚬⚬○○?

(b) Subtracting 7 from $7 + 19w$?

(c) Removing ⚬⚬⚬⚬ from ⚬⚬⚬⚬○○○○○?

(d) Adding $9x$ to $15 + 4x$?

(e) Doubling ⚬⚬○○○○○?

(f) Multiplying $6t + 9$ by 4?

(g) Removing half of ⚬⚬⚬⚬⚬⚬○○○○?

(h) Dividing $14r + 21$ by 7?

5 (a) Remove ○○○○ from both sides of

⚬○○○○=⚬⚬⚬○○○○○○○

(b) Remove ⚬⚬ from both sides of

⚬⚬○○○○○=⚬⚬⚬

6 (a) Subtract 5 from (both sides of) $3s + 5 = 4s + 7$.

(b) Multiply (both sides) of $2x + 4 = 3x + 1$ by 5.

(c) Divide (both sides) of $6m + 18 = 12m + 12$ by 6.

7 (a) What must be done to $3k + 5$ to get $3k$?

(b) What must be done to $3k$ to get k?

(c) What must be done to $5n + 17$ to get $5n$?

(d) What must be done to $5n$ to get n?

8 Solve these equations, picturing the work with bags and stones if necessary. Check your solutions by trying them in the original equations.

(a) $3a + 8 = 4a + 7$ (b) $2b + 6 = 3b + 4$

(c) $4c + 4 = 2c + 10$ (d) $3d + 9 = d + 13$

(e) $5e + 3 = 2e + 12$ (f) $6f + 3 = f + 13$

(g) $g + 9 = 5g + 6$ (h) $2h + 11 = 5h + 6$

9 (a) What is the result of adding 4 to $2y - 4$?

(b) What is the result of adding $2\frac{1}{2}$ to $3z - 2\frac{1}{2}$?

(c) What is the result of adding 7 to $3k - 2$?

(d) What is the result of adding 8 to $7\frac{1}{2}p - 4$?

10 (a) What must be added to $10x - 16$ to get $10x$?

(b) What must be added to $4x - 19$ to get $4x$?

11 Solve these equations, and check your answers.

(a) $4i - 8 = 3i - 7$ (b) $2j - 3 = 3j - 5$

(c) $3k - 3 = 4k - 7$ (d) $3y - 4 = 2y - 7$

(e) $2m - 5 = 4m - 13$ (f) $n - 5 = 3n - 3$

(g) $2p - 7 = 5p - 14$ (h) $2q - 7 = 5q - 5$

12 (a) What is the result of dividing $5y = 7$ by 5?

(b) What is the result of dividing $-2z = 3$ by -2?

13 Solve these equations and check your answers.

(a) $4r - 11 = 3r + 3$ (b) $3s - 4 = 4s + 7$

(c) $2t - 3 = 4t + 9$ (d) $5u - 3 = 2u + 6$

(e) $v - 6 = 3v + 2$ (f) $6w - 4 = w + 7$

14 In solving equations one often meets expressions with parentheses in them. These can be simplified by using the idea from Prob. 5 of Exercises 6.2. For example, we can simplify $2 - (x + 3)$ as $2 - x - 3$, which in turn may be simplified to $5 - x$. Similarly, we can simplify $5t - (4 - 2t)$ as $5t - 4 + 2t = 7t - 4$.

Simplify:

(a) $17 - (2 + 5x)$ (b) $2y - (5y - 11)$ (c) $-42z - (25 - 13z)$

9.3 **MORE ABOUT EQUATIONS**

Equations which are basically simple sometimes appear in disguised forms which make them look complicated at first. Here we consider a few of these.

EXAMPLE 1   SOLVE $3 = \frac{7}{10}(x + 2)$

A first step in isolating x would be to isolate $x + 2$, a number which is multiplied by $\frac{7}{10}$ in the given equation. To undo this multiplication we divide (both sides!) by $\frac{7}{10}$.

$$\frac{3}{\frac{7}{10}} = \frac{\frac{7}{10}(x + 2)}{\frac{7}{10}}$$

On the right division by $\frac{7}{10}$ undoes the multiplication, leaving $x + 2$.

$$3 \cdot \frac{10}{7} = x + 2$$

On the left $3 \div \frac{7}{10} = 3 \cdot \frac{10}{7}$, or $4\frac{2}{7}$.

$$4\frac{2}{7} = x + 2$$

Subtracting 2 completes the solution

$$\frac{-2 = -2}{2\frac{2}{7} = x}$$

EXAMPLE 2   SOLVE $\dfrac{3}{x + 2} = \dfrac{7}{10}$

Here the number to be found is in the denominator, where 3 is divided by $x + 2$. Multiplying by $x + 2$ undoes the division, leaving This equation was solved in Example 1.

$$3 = \frac{7}{10}(x + 2)$$

EXAMPLE 3   SOLVE $4 - (2z - 1) = 7$

The parentheses show that $2z - 1$, treated as a single number, is subtracted from 4 on the left. To undo this, we add $2z - 1$ to both sides.

$$\frac{+ (2z - 1) = + (2z - 1)}{4 \qquad\qquad 7 + (2z - 1)}$$

To isolate z, we first isolate $2z - 1$, which has 7 added to it, by subtracting 7.

$$\frac{- 7 = -7}{4 - 7 = 7 - 7 + (2z - 1)}$$

Now $4 - 7 = -3$

$$-3 = 0 + (2z - 1)$$

The parentheses on the right are no longer needed, so we have

$$-3 = 2z - 1$$

Adding 1

$$\frac{+1 \qquad\qquad +1}{}$$

gives

$$\frac{-2 = 2z}{2 \qquad 2}$$

Divide by 2 and simplify to get

$$-1 = z$$

EXAMPLE 4   SOLVE $\dfrac{3}{y} + \dfrac{5}{4y} = \dfrac{1}{2}$

Again the number we seek is in the denominator, this time in two fractions. We could begin by adding these two fractions (common denominator $4y$), but we shall simply multiply by $4y$.

$$4y\left(\frac{3}{y} + \frac{5}{4y}\right) = \frac{1}{2} \cdot 4y$$

Using the distributive property, we can rewrite the left to get

$$4y \cdot \frac{3}{y} + 4y \cdot \frac{5}{4y} = \frac{1}{2} \cdot 4y$$

Simplifying yields

$$4 \cdot 3 + 5 = 2y$$

or

$$17 = 2y$$

Dividing by 2 gives

$$\frac{17}{2} = y$$

These examples show that competence in arithmetic is essential for anyone who intends to learn algebra. A student enrolled in algebra despite inadequate preparation is condemned to almost certain frustration and failure, but to one who understands arithmetic, skill in solving simple equations is largely a matter of practice.

Exercises 9.3

1 (a) Multiply $\dfrac{1}{x}$ by x

    (b) Multiply $\dfrac{2}{y}$ by y

    (c) Multiply $\dfrac{z+1}{4}$ by 4

    (d) Multiply $\dfrac{2}{w+3}$ by $w+3$

    (e) Divide $2(a+17)$ by $a+17$

    (f) Divide $\dfrac{4x+7}{3}$ by $4x+7$

    (g) Add $4-t$ to $16-(4-t)$

    (h) Add x^2 to $3-x^2$

    (i) Subtract $2x+3$ from $42x+2x+3$

    (j) Multiply $\dfrac{2}{j+3}$ by $j+3$

2 (a) What must be done to $\dfrac{3}{a+1}$ to get 3?

    (b) What must be done to $4x(2x-7)$ to get $4x$?

    (c) What must be done to $4(2x-7)$ to get $2x-7$?

    (d) What must be done to $3d-14$ to get $3d$?

3 Solve these equations and check your solutions.

    (a) $\dfrac{4}{a+1}=2$

    (b) $\dfrac{4}{b-2}=5$

    (c) $7=\dfrac{3}{2-c}$

    (d) $-2=\dfrac{4}{d}$

    (e) $5-(e+2)=7$

    (f) $4f-3=6-(f+4)$

    (g) $8-(2+g)=14-(g+7)$

    (h) $4h-\left(2\frac{1}{2}-h\right)=19-\left(3h-\frac{1}{2}\right)$

    (i) $\dfrac{14}{i}=\dfrac{9}{2i}$ (careful!)

    (j) $\dfrac{1}{j+1}=\dfrac{3}{j+1}+1$

9.4 WORD PROBLEMS

"If I subtract from the double of my present age the treble of my age six years ago, the result is my present age. What is my age?" This problem from Chrystal's "Textbook of Algebra" sounds confusing in English, but it is easily solved with equations. The speaker's age is a number of years which we do not know and must find; we shall denote it by y. The other important numbers in the problem are the double of the speaker's age, represented as $2y$, and the treble (triple) of his age 6 years ago. Since 6 years ago his age was $y-6$, the treble of this is $3(y-6)$. Numerically, the problem states that if $3(y-6)$ is subtracted from $2y$, the result is y. As an equation, $2y-3(y-6)=y$. This is merely a shorthand restatement of the problem, stripped of (irrelevant) context. Equations, unlike English sentences, are subject to algebraic methods, which we now use to solve

$$2y - 3(y - 6) = y$$

Subtracting y $y - 3(y - 6) = 0$

Adding $3(y - 6)$ $y = 3(y - 6)$

Multiplying out the right (which does not change it) $y = 3y - 18$

Subtracting y $0 = 2y - 18$

Adding 18 $18 = 2y$

Dividing by 2 $9 = y$

It is easy to check that 9 is indeed a solution of the original equation. The last step is to translate the result back into verbal form. In this case it is easy; the speaker is a (precocious) nine-year-old.

The method used here is extremely powerful. It has three steps:

1. Express the problem in equations.
2. Solve the equations.
3. Interpret the solutions in terms of the original problem.

For beginners the most difficult part is usually the first step, which involves disregarding all information which is not numerical and translating the numerical information into equations. Skill at this requires practice. (The teacher must be careful here. Some students with reading problems find this hard, even though they do well in abstract mathematics. A poor reader has trouble enough without making his mathematics depend on his reading.) The following exercises in translation will help before we go on to more difficult problems. For simplicity, they begin with phrases instead of sentences.

Exercises 9.4

1   Let j stand for the number of apples Joe picked. Express in algebra:
   (a) Twice as many apples as Joe picked.
   (b) Ten more apples than Joe picked.
   (c) Fifteen fewer apples than Joe picked.
   (d) Three fewer than twice as many as Joe picked.
   (e) Seven and a half times as many apples as Joe picked.
   (f) Five more than four times as many apples as Joe picked.
   (g) The number of apples Joe picked subtracted from 100.
2   Bill earns d dollars a year, paid at a steady rate throughout the year.
   (a) How much does he earn at that rate in 4 years?
   (b) How much does he earn at that rate in 7 years?
   (c) How much does he earn per month?
   (d) How much is $200 more than 3 years of his wages?
3   There are c children in this class, and I have marshmallows to give them. Express the number I have if:

(*a*) There are just enough to give out one to each child.

(*b*) There are just enough to give three to each child.

(*c*) There are just enough to give two to each child and a third one to seven of the children.

(*d*) There are three fewer than would be necessary to give out five to each child.

(*e*) I had 200 marshmallows, but I have already given out three to each child.

4 Suppose *A* is my present age in years.

(*a*) What will my age be fifteen years from now?

(*b*) How old will I be when I am twice my present age?

(*c*) How old will I be three years after I double my present age?

(*d*) Write an equation to express the statement: "In fifteen years I will be three years older than twice my present age."

(*e*) Solve the equation to find my present age.

5 Candies cost *c* cents a piece.

(*a*) How much do seven candies cost?

(*b*) If I have one cent more than enough to buy seven candies, express how much money I have.

(*b*) If I have one cent more than enough to buy candies, express how much money I have.

(*c*) How much money do ten candies cost?

(*d*) If I lack twenty cents of being able to buy ten candies, how much money have I?

(*e*) How much do the candies cost apiece if I have one cent more than enough to buy seven candies but lack twenty cents of being able to buy ten candies? (Express this as an equation and solve it.)

6 I took *x* dollars with me to the races.

(*a*) On the first race I doubled my money. How much did I have then?

(*b*) On the second race I lost $10. How much did I have then?

(*c*) At that point I had $34 left. How much did I begin with?

7 Farmer Jones has *j* cows, and his neighbor farmer Smith has twice as many. Farmer Brown has five fewer cows than Jones and Smith together, and all three together have eighty-five cows. How many cows have Jones, Smith, and Brown? (Express how many cows Brown and Smith have in terms of *j*; then make up an equation which states that the three together have eighty-five cows.)

8 Sarah needs *y* yards of cloth for curtains, and her friend Adzemantha needs twice as many yards of the same material. If they buy a remnant (which is on sale) of cloth 25 yards long, they will have 3 yards left over after each takes what she needs. How many yards of cloth does Sarah need?

9 On Tuesday a certain record shop sold 5 fewer records than it had on Monday, and on Wednesday it sold 12 more than it did on Monday. The total sales for Thursday and Friday were 17 more than the total for Monday and Wednesday. If the sales for the entire period totaled 141 records, how many were sold on Monday? *Hint*: Let *r* stand for the number of records sold on Monday. Express the number of records sold on subsequent days in terms of *r* and make an equation which relates *r* to the total sales for the week.

10 In the class I visited, half the children were at work on their English themes, and a third of the rest were engaged in a science project. The remaining eight students were doing arithmetic. How many were in the class? (Let *c* = the number of children in the class.)

11 A certain basketball player scored 3 more points in his second game than in his first, but in the third game he scored only half as many as he did in the second game. He scored a total of 52 points in the three games. How many did he score in each?

12 A man spends one-third of his money and loses two-thirds of the remainder. He then has 12 pieces. How much money had he at first? (From a text of 1484 by Nicolas Chuquet.)

13 Half the length of this strip of cloth is colored red, and one-fifth of it is blue. Half the remainder is green, and the other 2 feet are yellow. How many feet long is the strip of cloth?

14 I am thinking of a number. If I add 5 to my number and double the result, I get the same number as if I subtracted 1 from the number and tripled the result. What is the number?

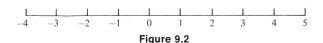

Figure 9.2

15 A woman bought some apples, oranges, and pears at the market. She bought twice as many oranges as pears, and $4\frac{1}{2}$ times as many pears as apples. In all she bought 87 pieces of fruit. How many did she buy of each?

16 If a certain number is multiplied by 7, the result is 20 less than the result of tripling the number. What is the number?

9.5 INEQUALITIES

Equations are not the only useful numerical statements. Often it is convenient to be able to compare numbers which are not equal, and for this *inequalities* are used. An inequality is simply a statement that one number is larger or smaller than another.

A number n is said to be *greater* than a number m if $n - m$ is positive or, what amounts to the same thing, if n lies to the right of m on the number line in Fig. 9.2. In this case m is also said to be *less* than n. For example, 2 is greater than -5 or, equivalently, -5 is less than 2. Inequalities are written in much the same way as equations except that the equals sign is replaced by either $<$, which means "is less than," or $>$, which means "is greater than." These signs are akin to the *crescendo* (increasing) and *decrescendo* (decreasing) signs in music. The pointed end belongs by the lesser number (or softer tone), and the growing width of the sign shows that the number (or loudness) increases toward its open end. Our example above may be expressed as $2 > -5$ or $-5 < 2$.

Inequalities, like equations, may be true, false, or conditional. A number which satisfies (makes true) a conditional inequality is called a *solution* of it, and an inequality is considered solved when all its solutions have been found. The basic technique for solving inequalities is similar to that for solving equations, but in practice it is often more awkward to use. For our purposes it is not necessary to be able to solve inequalities, but it is important to be familiar with the signs $<$ and $>$ and their meanings, for they are widely used.

Exercises 9.5

1 Which, if any, of the following inequalities are true?

(*a*) $5 < 4$ (*b*) $-10 < -2$ (*c*) $-2 < 10$ (*d*) $0 < -5$
(*e*) $\frac{1}{2} < -4$ (*f*) $-2 > -6$ (*g*) $3^2 > 2^2$ (*h*) $(-4)^2 < (-2)^2$
(*i*) $(-5)^2 > 0$ (*j*) $36 > 6^2$

2 Find two numbers m and n to show that it is possible for m to be less than n and yet m^2 to be greater than n^2.

3 Often in books one sees the signs $\leqslant$ and $\geqslant$. These are combinations of equals signs with the signs for inequalities and are read "less than or equal to" and "greater than or equal to," respectively. For example, $x > 2$ is true for numbers greater than 2 but not for 2 itself, but $x \geqslant 2$ is true for 2 itself as well as greater numbers. Which, if any, of these are true?

(a) $0 \leqslant 7$ (b) $-2 \leqslant 4$ (c) $-2 \geqslant -1$

(d) $3^2 \leqslant 9$ (e) $(-3)^2 \geqslant 3^2$ (f) $-3^2 \leqslant 0$

CHAPTER 10
THE DECIMAL SYSTEM

The concept of number has been refined and expanded over the centuries, and so have ways of expressing numbers in writing. Our modern system is now so perfected that it is often taken for granted. Here we consider the decimal system of numeration in some detail.

10.1 NUMBERS AND NUMERALS

Numbers are abstract ideas. Like any abstract idea, such as truth or beauty, they are not easy to define accurately. We do not need a careful definition of numbers here (an intuitive idea will do), but we shall need to distinguish carefully between numbers themselves and written symbols which represent them. Such symbols are called *numerals*. In daily life the distinction between ideas and symbols which represent them is often blurred. For example, what is in this box? LEMONADE . The box is not wet. Instead of lemonade it contains printed marks which we interpret as representing the drink. Likewise 4 is just a printed mark, but we interpret it as representing a number, so it is a numeral. A few other numerals which stand for the same number as 4 are: $\frac{8}{2}$, 2^2, $-2 + 6$, ● ● ● ● (ancient Mayan), IV (Roman), and δ (Greek). In most circumstances it is tedious to labor the distinction between number and numeral; we have ignored it in previous chapters, but we need it here.

10.2 THE DECIMAL SYSTEM

The decimal system originated in India between 300 B.C. and A.D. 800, though the details are not known today. It spread throughout the Arab world and eventually to Europe, where the "Liber Abaci" by Fibonacci did much to popularize it.[1] Until the decimal system arrived, Europeans used Roman numerals. These are so clumsy that the few people who could do even simple arithmetic without an abacus were often suspected of supernatural powers. The clever new system made it possible to replace the abacus with a few techniques anyone could learn.

In the decimal system any number is expressed in terms of just 10 symbols or *digits*, 0, 1, 2, 3, 4, 5, 6, 7, 8, and 9. The meaning of a digit depends on where it is written. For example, the 8s in 830 and 48 have different meanings. One who understands the system realizes that page 301 is nearer page 299 than it is to 103 or 401, though the digits of 103 and 401 look more like those of 301 than do those of 299. Locating a page by number is like looking up a word in a dictionary. The leftmost digit, like the first letter of the word, gives us the first clue. Successive digits (letters) to the right narrow the search until the page (word) is found. This is the pattern of decimal numeration.

How is the meaning of each digit determined? This depends on certain factors which are not written explicitly but are implied by the positions of the digits. For example, 27,458 written to show the hidden factors is $2 \cdot 10,000 + 7 \cdot 1,000 + 4 \cdot 100 + 5 \cdot 10 + 8 \cdot 1$. The right-hand digit of an integer is called the *units* digit, since its hidden factor is 1. Similarly, other digits are called the *tens digit, hundreds digit,* and so on, after their hidden factors. The pattern of the hidden factors (Fig. 10.1) may be extended upward to higher powers of 10. It may also be extended to smaller numbers by continuing to divide by 10. The resulting list of hidden factors is

$$\cdots 10,000 \quad 1,000 \quad 100 \quad 10 \quad 1 \quad \frac{1}{10} \quad \frac{1}{100} \quad \frac{1}{1,000} \quad \frac{1}{10,000} \cdots$$

where the dots indicate that the list extends without end in both directions. Unless the number expressed is an integer, a *decimal point* is put just to the right of the units digit as a reference marker. For example, 24.79 means

$$2 \cdot 10 + 4 \cdot 1 + 7 \cdot \frac{1}{10} + 9 \cdot \frac{1}{100}$$

For algebraic work, the form a/b is very convenient, but for purely numerical computation the decimal notation has great advantages. Our methods for computing with integers carry over directly to decimal fractions, since these methods result directly from decimal numeration. In particular, decimal fractions can be added just like integers, avoiding the worry about common denominators. Fractions are also easier to compare in decimal form than expressed as a/b. For example, which is larger, $\frac{29}{91}$ or $\frac{85}{272}$? You could tell

---

[1]Leonardo Pisano (1170–1250), known as Fibonacci, was born in Pisa, grew up in North Africa, and traveled to Egypt, Greece, Sicily, and Syria. His "Liber Abaci," first published in 1202, explained and advocated Hindu-Arabic numeration. It also contained a problem involving what are now called Fibonacci numbers, 1, 1, 2, 3, 5, 8, . . . (see Exercises 5.1, Prob. 13).

To move down,   10,000 = 10⁴   To move up
a step, divide   1,000 = 10³   a step,
by 10.   100 = 10²   multiply by 10.
    10 = 10¹
    1 = 10⁰   (as we shall see later)

Figure 10.1

at a glance if they were in decimal form ($\frac{29}{91}$ is about 0.319, while $\frac{85}{272}$ is roughly 0.313).

Exercises 10.2

1 Label the marked points with decimal fractions.

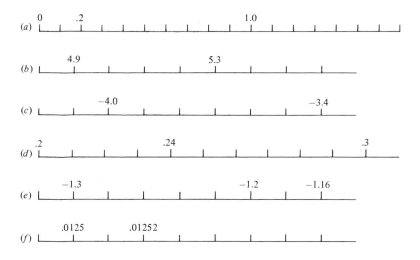

2 If you feel you need practice adding and subtracting decimal fractions, try these box puzzles (or make some up for yourself). Not only are they self-checking, but you can check them a second way by rewriting the fractions in the form x/y.

(a)

.03	3.1	
.17	2.9	

(b)

1.8		2.8
	3.7	
		8.2

(c)

3.07		
	2.03	
1.96		4.10

3 Approximations are used for convenience or because precise numbers are not known. For example, in 1970 the federal government spent $958,260,537 in Hawaii. It may be convenient to approximate this as $958,000,000 ($958 million) or, more crudely, as $1,000,000,000 ($1 billion). This kind of approximation is called *rounding off*. Round off:

(*a*) 407 to the nearest hundred
(*b*) 392 to the nearest hundred
(*c*) 14,878,294 to the nearest thousand
(*d*) 14,878,294 to the nearest million
(*e*) 42.038 to the nearest integer
(*f*) 42.038 to the nearest tenth
(*g*) 42.038 to the nearest hundredth
(*h*) −215.507 to the nearest integer

4 Placing the decimal point in a product confuses many people. There is a rule for this which many use but few understand: place as many digits to the right of the decimal point in the product as there are digits to the right of the decimal points in all the factors. We shall consider this more carefully in Exercises 12.3. For now, here are other ways to place the decimal point.

(*a*) Estimation; for example, 4.1×3.07. Ignoring decimal points, we have $41 \times 307 = 12,587$. Where does the decimal point belong? Rounded off to the nearest integers, 4.1 and 3.07 are ____ and ____. Therefore the product is roughly ____, so $4.1 \times 3.07 =$ ____.

(*b*) $4.1 \times 3.07 = \dfrac{}{10} \cdot \dfrac{}{100} = \dfrac{}{1,000} = 12 \frac{587}{1000} = 12.587$

If you feel you need practice with these ideas, try them out on a few examples of your own.

5 Division of one decimal fraction by another poses no new problems. For example,

$$14.071 \div 0.45 = \frac{14,071}{1,000} \div \frac{45}{100} = \frac{14,071}{1,000} \cdot \frac{100}{45}$$

$$= \frac{14,071}{1,000} \cdot \frac{1,000}{450} = \frac{14,071}{450}$$

Now carry out the division of integers. If you feel you need practice, try these, checking your results by multiplying.

(*a*) $84 \div 2.3$ (*b*) $142.5 \div 0.031$

6 Can you reconstruct this box puzzle with border numbers?

	2.0		3.2
1.7			
−4.6			

10.3 DECIMALS AND FRACTIONS

How can a fraction x/y be expressed as a decimal? For example, what is the decimal form of $\frac{35}{16}$? Since $\frac{35}{16}$ means $35 \div 16$, we write 35 as 35.0000 (we could put more 0s if needed) and divide beyond the decimal point, as in Fig. 10.2. (Note that here it is convenient to use the streamlined form of the division algorithm instead of the introductory form presented in Sec. 6.5.) Is it really true that $2.1875 = \frac{35}{16}$? How could you check it? Would the same process work as well with other fractions?

```
   2.1875              .3333
16 )35.0000        3 )1.0000
   32                  9
   ──                  ──
   30                  10
   16                   9
   ──                  ──
   140                 10
   128                  9
   ──                  ──
   120                 10
   112                  9
   ──                  ──
    80                  1
    80
   ──
```

Figure 10.2   Figure 10.3

Indeed it can lead to some surprises. What if we use it to express $\frac{1}{3}$ as a decimal? We write 1 as 1.0000. The work appears in Fig. 10.3. Note the process has not come to an end. Will it ever end? Clearly not, for each step is like the one before. Thus the decimal form of $\frac{1}{3}$ is 0.333 $\cdots$, where the dots indicate that the 3s continue endlessly. Such an unending decimal is called an *infinite* decimal. Those with only a finite number of digits are called *finite* decimals.

Exercises 10.3

1 Check that $2.1875 = \frac{35}{16}$:
  (a) By multiplying 2.1875 by 16
  (b) By reducing $\frac{21,875}{10,000}$ to lowest terms

2 (a) In sports a team's *winning percentage* is the fraction of its games which it wins expressed as a decimal rounded off to three places. For example a team which wins 7 games out of 15 has a winning percentage of .467, since $\frac{7}{15} = .4666 \cdots$. In 1906 the Chicago Cubs won 116 games and lost only 36. What was their winning percentage?
  (b) A baseball player's *batting average* is the fraction of his official times at bat in which he made hits. This is usually expressed as a decimal, rounded off to three places. For example, a player who makes 100 hits in 350 times at bat has a batting average of .286, since $\frac{100}{350} = .285714 \cdots$. In 1968 Pete Rose batted .355 and came to bat 622 times. How many hits did he make?

3 *Percent* means, literally, *per hundred*. For example, 50 percent is $\frac{50}{100}$, and 67 percent is $\frac{67}{100}$. What percent is:
  (a) $\frac{3}{5}$ (b) $\frac{17}{20}$ (c) $\frac{50}{93}$

4 Advertisements claim a product to be $99\frac{44}{100}\%$ pure. What percentage of impurities in this product is implied by that claim?

5 A tire dealer advertises that anyone who buys three tires at list price can buy a fourth at half price. If you buy four tires from this dealer, what percentage of the list price do you save?

6 (a) A store announced a 10 percent price reduction on some merchandise. When this did not stimulate sales sufficiently, prices were cut by an additional 15 percent. By what percentage were prices reduced overall? *Hint*: The answer is not 25 percent; the 15 percent discount was applied to an already slashed price.
  (b) Another store made the same price reductions in reverse order, first cutting prices by 15 percent and then reducing prices by another 10 percent (of its already reduced prices). By what percentage did it reduce prices overall?
  (c) Compare your answers from parts (a) and (b). What general principle does this illustrate?

7 If a candidate wins 457,231 votes out of 802,377 cast, what percentage of the votes did he get? (Round off your answer to the nearest whole number.)

8 (*a*) How many cents are *n* nickels worth?

(*b*) How many cents are *d* dimes worth?

(*c*) I have three times as many nickels as dimes. Altogether my nickels and dimes are worth $2.25. How many of each have I?

9 (*a*) Express $\frac{2}{3}$ as a decimal by dividing.

(*b*) Double the decimal for $\frac{1}{3}$. Does the result agree with part (*a*)?

(*c*) Triple the decimal for $\frac{1}{3}$. You should get a decimal for $\frac{3}{3} = 1$. Are you surprised at the result? We shall return to it later.

10 (*a*) Find the decimal for $\frac{1}{7}$, carrying the division far enough to see the pattern. (You need at least 7 digits, maybe as many as 12.)

(*b*) Fill in the blanks to fit the patterns.

$$7 \cdot 2 = \underline{1} \quad \underline{4}$$
$$\underline{\quad} \cdot 4 = \qquad \underline{2} \quad \underline{8}$$
$$7 \cdot 8 = \qquad\qquad \underline{5} \quad \underline{\quad}$$
$$7 \cdot 16 = \qquad\qquad\quad \underline{1} \quad \underline{\quad} \quad \underline{\quad}$$
$$7 \cdot \underline{\quad} = \qquad\qquad\qquad\qquad \underline{\quad} \quad \underline{\quad} \quad \underline{4}$$
$$7 \cdot 64 = \qquad\qquad\qquad\qquad\quad \underline{\quad} \quad \underline{\quad} \quad \underline{\quad}$$

Add digits in each column: $\underline{\quad}\ \underline{\quad}\ \underline{\quad}\ \underline{\quad}\ \underline{\quad}\ \underline{\quad}\ \underline{\quad}\ \underline{\quad}\ \underline{\quad}\ \underline{\quad}\ \underline{\quad}$

(*c*) How are parts (*a*) and (*b*) related?

(*d*) By any means convenient, find the decimals for $\frac{2}{7}, \frac{3}{7}, \frac{4}{7}, \frac{5}{7}$ and $\frac{6}{7}$ to at least six digits. What curious patterns do you notice in your answers?

(*e*) Adding the decimals for $\frac{1}{7}$ and $\frac{6}{7}$ should yield a decimal for $\frac{7}{7} = 1$. Try it. Is the result surprising? What of $\frac{2}{7} + \frac{5}{7}$ and $\frac{3}{7} + \frac{4}{7}$? We shall return to this later.

11 (*a*) Find the decimal for $\frac{1}{14}$ by division.

(*b*) Check your result from part (*a*) by the fact that $\frac{1}{14} + \frac{1}{14} = \frac{1}{7}$.

(*c*) Find the decimal for $\frac{10}{14}$ by dividing 10 by 14.

(*d*) How are your answers for (*a*) and (*c*) related?

(*e*) As another check, observe that you already found the decimal for $\frac{10}{14}$ in part (*d*) of Prob. 10.

*10.4 MORE ABOUT DECIMALS

By now you may have noticed some curious things about the decimal forms of fractions. Some of them, like $\frac{1}{5} = 0.2$ and $\frac{1}{8} = 0.125$, are just a few digits long, while others go on forever. Those which have infinitely many digits seem to have repeating patterns of digits such as 142857 (which repeats endlessly in the decimal for $\frac{1}{7}$) or 3 (which repeats forever in the decimal for $\frac{1}{3}$). Which fractions have finite decimal forms? If the decimal form of a fraction goes on forever, does it necessarily have a repeating pattern of digits? If so, why? Can you tell ahead of time how many digits long the repeating pattern of digits will be? If you are given an infinitely long decimal which represents a fraction, how can you find the fraction it represents? The following exercises are designed to help you answer these questions. For convenience, we shall work only with fractions whose numerators are 1; once you understand them, you can easily extend the work to other cases.

Exercises 10.4

1 (a) To begin, compile some evidence by classifying each of these fractions according as its decimal form is or is not infinitely long:

$$\frac{1}{2} \quad \frac{1}{3} \quad \frac{1}{4} \quad \frac{1}{5} \quad \frac{1}{6} \quad \frac{1}{7} \quad \frac{1}{8} \quad \frac{1}{9} \quad \frac{1}{10} \quad \frac{1}{11} \quad \frac{1}{12} \quad \frac{1}{13} \quad \frac{1}{14} \quad \frac{1}{15} \quad \frac{1}{16}$$

(Many of these have been expressed as decimals in the preceding parts of this chapter and exercises, and $\frac{1}{13}$ is shown in Prob. 2 on p. 102, so only a few remain for you to divide.)

(b) What distinguishes those fractions whose decimals are finite? To see, factor their denominators into primes. You will find that only certain primes are involved. Which are they?

(c) Use part (b) to predict which of the following fractions have finite decimal forms:

$$\frac{1}{20} \quad \frac{1}{30} \quad \frac{1}{32} \quad \frac{1}{40} \quad \frac{1}{45} \quad \frac{1}{55} \quad \frac{1}{60} \quad \frac{1}{65} \quad \frac{1}{75} \quad \frac{1}{120} \quad \frac{1}{125}$$

(d) Check your predictions without carrying out the division by reasoning this way. A finite decimal can be expressed with a denominator which is a power of 10, for example, $0.0625 = 625/10^4$. Reducing this to lowest terms cannot introduce any new prime factors in the denominator, so if a fraction has a finite decimal form, its denominator (assuming the fraction is in lowest terms) can have no prime factors other than the prime factors of 10, namely 2 and 5.

(e) If a fraction $1/n$ has a finite decimal expression, can you predict how many digits long it will be? You may be able to reason directly from part (d). Otherwise begin by gathering data, completing the table, and looking for patterns in it. (Here we use the fact that $2^0 = 1$ and $5^0 = 1$, a point considered more carefully in Chap. 12.) When you have filled in the table, look for a relationship between the number of digits in the decimal form and the exponents in the prime factorization of the denominators. Can you see why the pattern occurs? *Hint*:

$$\frac{1}{16} = \frac{1}{2^4} = \frac{5^4}{2^4 \cdot 5^4} = \frac{5^4}{10^4} = \frac{625}{10,000} = 0.0625$$

	$\frac{1}{2} = \frac{1}{2^1 \cdot 5^0} = 0.5$	$\frac{1}{4} = \frac{1}{2^2 \cdot 5^0} = 0.25$	$\frac{1}{8} = \frac{1}{2^3 \cdot 5^0} = 0.125$
$\frac{1}{5} = \frac{1}{2^0 \cdot 5^1} = 0.2$	$\frac{1}{10} = \frac{1}{2^1 \cdot 5^1} = 0.1$	$\frac{1}{20} = \frac{1}{2^2 \cdot 5^1} = 0.05$	$\frac{1}{40} = \frac{1}{2^3 \cdot 5^1} =$
$\frac{1}{25} = \frac{1}{2^0 \cdot 5^2} =$	$\frac{1}{50} = \frac{1}{2^1 \cdot 5^2} =$	$\frac{1}{100} = \frac{1}{} = 0.01$	$\frac{1}{200} = \frac{1}{} = 0.005$
$\frac{1}{125} = \frac{1}{2^0 \cdot 5^3} =$	$\frac{1}{250} = \frac{1}{2^1 \cdot 5^3} =$	$\frac{1}{500} =$	$\frac{1}{1,000} =$

2 If the decimal form of a fraction is infinite, does it necessarily have a repeating pattern of digits? If so, how long is the pattern? To investigate this, we consider the example of $\frac{1}{13}$ in detail. To express $\frac{1}{13}$ as a decimal we must divide 1 by 13. This in turn involves subproblems as shown here.

$$
\begin{array}{r}
.076923 \\
13\overline{)1.000000 \cdots} \\
\end{array}
$$

Subproblem	**1**	$10 \div 13$	91
,,	**2**	$100 \div 13$	90
,,	**3**	$90 \div 13$	78
,,	**4**	$120 \div 13$	120
,,	**5**	$30 \div 13$	117
,,	**6**	$40 \div 13$	30
,,	**7**	Same as subproblem 1	26

91
90
78
120
117
30
26
40
39
10

(*a*) What can you say about subproblems 8 to 13?

(*b*) What does this tell you about the decimal for $\frac{1}{13}$?

(*c*) Each subproblem involves dividing 13 into a number formed by adjoining a zero to the remainder from the previous subproblem. Dividing by 13 can result in only 1 of 13 remainders: 0, 1, 2, 3, 4, 5, 6, 7, 8, 9, 10, 11, 12. In this case, however, the remainder 0 cannot occur, as if it did the decimal for $\frac{1}{13}$ would end and we know from Prob. 1 that it does not end. This guarantees that the repeating pattern of digits in the decimal for $\frac{1}{13}$ is no more than 12 digits long (though in fact it is only 6). Applying this reasoning to other fractions, you can guarantee that the repeating pattern of digits in the decimal for:

$\frac{1}{17}$ is no more than _____ digits long.

$\frac{1}{18}$ is no more than _____ digits long.

$\frac{1}{19}$ is no more than _____ digits long.

3 Problem 2 probably convinced you that if a fraction $1/n$ has an infinite decimal form, it involves a periodically repeating cycle of at most $n-1$ digits. However, you may have noticed that in certain cases this repeating pattern does not begin right away. For example, the decimal for $\frac{1}{12}$ begins its endless repetition of 3s only after two preliminary digits (0 and 8), and the decimal for $\frac{1}{15}$ waits for an initial 0 before beginning its endless sequence of 6s.

(*a*) Can you find other examples like these? (You met two in Prob. 1.)

(*b*) Can you find a pattern in your examples which allows you to predict how many digits a decimal goes before settling into a cyclic pattern? *Hint*: this pattern is closely related to the one in part (*e*) of Prob. 1.

4 Let us summarize the results of the previous problems about the decimal for $1/n$.

  1. If n has no prime factors other than 2 and 5, so we can write $n = 2^r \cdot 5^s$, then $1/n$ has a decimal form which ends after a number of digits equal to the larger of r and s.

  2. If n has no prime factors of 2 and 5, the decimal for $1/n$ consists of an infinitely repeating cycle of at most $n-1$ digits.

  3. If n has prime factors of 2 or 5 as well as other primes, we can write $n = 2^r 5^s m$, where m is not a multiple of 2 or 5. Then $1/n$ has a decimal form which consists of a preliminary sequence of digits (equal in number to the larger of r and s) followed by an endlessly repeating cycle of at most $n-1$ digits. (In fact, as you may suspect, this cycle is actually no more than $m-1$ digits long.) Is there a more precise rule for the length of the digital cycles in cases 2 and 3?

Yes, but it is not one you would be likely to guess. If we write n as $2^r 5^s m$, where in case 2 we take $r = s = 0$, then the length of the periodic cycle of digits in the decimal for $1/n$ is the smallest power to which 10 can be raised to get a number which is 1 more than a multiple of m. For example, in the decimal for $\frac{1}{56}$, we write $56 = 2^3 5^0 7$. The larger (3) of the exponents 3 and 0 tells us that the decimal for $\frac{1}{56}$ waits 3 digits before beginning its periodic behavior, and the fact that 10^6 is the smallest power of 10 which is 1 more than a multiple of 7 tells us that the repeating cycle is 6 digits long. Check this by computing the first 10 digits in the decimal for $\frac{1}{56}$.

5   Given an infinite periodic decimal, how do you find the fraction it represents? Here is a way to handle such problems. For concreteness we demonstrate with $0.373737 \cdots$. First let a letter, say N, stand for the number. Then observe the length of the cycle of digits, in this case 2 digits long. Multiply the equation $N = 0.373737 \cdots$ by 10^2 (the power of 10 is the length of the digital cycle) to get

$$100N = 37.373737 \cdots$$
Subtract $\quad\;\; N = \;\; 0.373737 \cdots$
$$99N = 37.000000 \cdots$$

Solve for N to get $N = \frac{37}{99}$. You can check that dividing 37 by 99 does indeed yield the infinite decimal we began with. Apply this process to find fractions equal to:

(a) $0.222 \cdots$

(b) $0.431431431 \cdots$

(c) $0.9999 \cdots$ [Does your answer surprise you? Compare with Prob. 9 of Exercises 10.3]

(d) $0.13252525 \cdots$ (repeats 25s forever)

CHAPTER 11
NONDECIMAL NUMERATION

The choice of ten as the base of our numeration is not a mathematical necessity. The reasons for it are mainly historical and no doubt come from counting on fingers. To barefoot people twenty is also a natural choice (for example, "fourscore and seven"), and some South American tribes base their numeration on five, in effect counting on the fingers of only one hand. Other nondecimal systems have been used. The ancient Babylonians based their numeration on sixty, and to them we owe the division of the hour into sixty minutes and the minute into sixty seconds. The users of nondecimal numeration are not all ancient or primitive. Computers do their internal arithmetic with numerals based on two or eight, and nondecimal numeration is used in schools as a way of reviewing "place value." Our brief study of nondecimal numeration will make you rethink some basic ideas and give you some idea of what a beginner faces as he struggles with the decimal system.

11.1 INTRODUCTION TO BASE FIVE

A tribe which uses the fingers of only one hand for counting counts something like this: "one, two, three, four, hand, hand and one, hand and two, hand and three, hand and four, two hands," and so on. If we picture the objects being counted by tally (itself a crude system of numeration), this is summed up in Fig. 11.1. In the right-hand column the various numbers are expressed in a system of numeration much like the decimal system, except that it uses only the digits 0, 1, 2, 3, and 4. This is called the *base-five* system of numeration. In this system the numeral 10 should be read "one zero" and not "ten" because it does not represent the number we usually call ten. We shall use words like "ten" and "eleven" for numbers, regardless of how they are written.

By following the patterns, can you extend the table in Fig. 11.1 to higher numbers? What is the largest number that can be expressed in base five as a

Tally	Spoken	Base-five numeral
I	one	1
II	two	2
III	three	3
IIII	four	4
︙	hand	10
︙ I	hand and one	11
︙ II	hand and two	12
︙ III	hand and three	13
︙ IIII	hand and four	14
︙ ︙	two hands	20
︙ ︙ I	two hands and one	21

Figure 11.1

Figure 11.2

two-digit numeral? Continuing to count in base five, the first number to need three digits will be written 100. What number does this represent? We shall call it a "hand of hands" and picture it like Fig. 11.2.

Exercises 11.1

1 What base-five numerals do these tally pictures represent?

(a) (b)

2 Draw tally pictures to represent these base-five numerals.
(a) 314   (b) 101   (c) 320
3 This number line was labeled with base-five numerals, but some were erased. Can you fill it in?

$$^-10 \qquad ^-1 \quad 0 \quad 1 \quad 2 \qquad 11$$

11.2 COMPUTATION IN BASE FIVE

Computation in base five may look strange at first, but the methods are essentially those used with base ten. To begin, we add 11 (hand and one) and 12 (hand and two), using tally pictures as a guide.

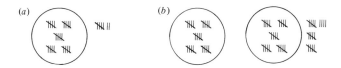

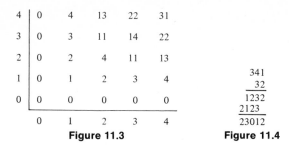

4	0	4	13	22	31
3	0	3	11	14	22
2	0	2	4	11	13
1	0	1	2	3	4
0	0	0	0	0	0
	0	1	2	3	4

Figure 11.3

$$\begin{array}{r} 341 \\ 32 \\ \hline 1232 \\ 2123 \\ \hline 23012 \end{array}$$

Figure 11.4

What base five numeral represents this sum? It *looks* as if we had added eleven and twelve in base ten, using the familiar columnwise addition.

But not all addition looks like base ten. For example, what if we add 12 (hand and two) and 14 (hand and four)?

Pictorially, these are ⲧⱨ II
and ⲧⱨ IIII

Adding pictorially yields ⲧⱨ ⲧⱨ IIIIII

As it stands, this does not correspond to a base-five numeral, but when it is regrouped as ⲧⱨ ⲧⱨ ⲧⱨ I , it corresponds to the base-five numeral 31. This, too, could have been found by ordinary columnwise addition, provided that (since we have no digit above 4) we carry fives instead of the tens we are used to. The regrouping step in our example corresponds to carrying. A child who grew up with base five would not find this unusual.

Addition problems may be longer than these, but in principle they are no more difficult. Subtraction is also straightforward, if one keeps in mind that borrowing involves fives instead of the tens we are used to.

What about multiplication? This, too, is much the same as with base ten, but one needs to know fewer single-digit multiplication "facts." These can be found by repeated addition, but for convenience they are given in Fig. 11.3, since you may find some of them strange at first sight.

As a sample multiplication with base five, Fig. 11.4 shows 341 · 32. Here the line 1232 represents 2 · 341 and was found this way: 2 · 1 = 2, which is the right-hand digit in 1232. Next 2 · 4 = 13, so we write the 3 and save the 1. Now 2 · 3 is 11, which, with the 1 which was saved, is 12. The second partial product 2123 is found similarly, and the two are added as discussed above.

Since multiplication is done one digit at a time, it helps to know the basic combinations shown in Fig. 11.3 just as it helps to know the basic facts in base ten. However, overemphasis on memorizing the table has been a serious stumbling blocks for many youngsters. Those who have trouble learning the multiplication table may not understand multiplication as repeated addition or may be weak at addition itself. The multiplication table has little meaning to such youngsters; drill in memorizing it at this stage can be counterproductive by fostering the feeling that mathematics is authoritarian, meaningless, and boring.

The following exercises are a sampler of base-five computation problems, arranged in increasing order of difficulty. Do only those which you feel will help you increase your understanding.

Exercises 11.2

1   What base-five additions to these pictures represent?

   (*a*)   卌 卌 |||
   　　　　　　卌 ||
   　　────────────
   　　卌 卌 卌 ||||| = 卌 卌 卌 卌

   (*b*)   卌 卌 卌 ||
   　　　　　卌 卌 ||||
   　　────────────── = (卌 卌 / 卌 / 卌 卌)   卌 |
   　　卌 卌 卌 卌 卌 ||||||

2   Add in base five, using tally pictures to guide and check your work.
   (*a*) 41 + 102   (*b*) 34 + 41   (*c*) 30 + 41   (*d*) 30 + 314
3   Experiment with box puzzles written in base five. Are their properties similar to those written in base ten?
4   What base-five subtractions do these tally pictures represent?

   (*a*) 卌 卌 ||| − 卌 | = 卌 || (*b*) (卌 卌 / 卌 / 卌 卌)   卌 卌 |||| − 卌 卌 卌 = 卌 卌 卌 ||||

5   Subtract in base five, using tally pictures to check and show regrouping as necessary.
   (*a*) 423 − 202   (*b*) 21 − 14   (*c*) 300 − 24
6   Fill in the missing parts of this base-five box puzzle.

	31	112
41		211

7   Check the multiplication table in Fig. 11.3 by carrying out the repeated additions. (Some combinations duplicate each other, and some are trivial, so only a few really need checking.)
8   Multiply and add to complete this base-five box puzzle with borders.

	2	4
3		
4		

9   Can you fill in the missing parts of this base-five box puzzle with borders?

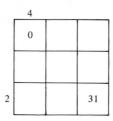

10   The numeral 23 represents prime numbers both in base five and in base ten. Can you find another two-digit numeral with this property?

11   Use the table in Fig. 11.3 to carry out these base-five divisions.

(*a*) 13 ÷ 4   (*b*) 22 ÷ 3   (*c*) 14 ÷ 3
(*d*) 11 ÷ 3   (*e*) 13 ÷ 2   (*f*) 31 ÷ 4

12   Carry out these base-five multiplications, and look for a pattern.

(*a*) 4 · 10   (*b*) 11 · 10   (*c*) 23 · 10
(*d*) 10 · 112   (*e*) 100 · 134   (*f*) 241 · 1000

13   Viewing division as repeated subtraction, carry out these base-five divisions. The patterns from Prob. 12 may help, as may Prob. 11.

(*a*) 130 ÷ 4   (*b*) 220 ÷ 3   (*c*) 1400 ÷ 3   (*d*) 1400 ÷ 30
(*e*) 11000 ÷ 30   (*f*) 13000 ÷ 400   (*g*) 134 ÷ 4   (*h*) 231 ÷ 3
(*i*) 332 ÷ 2   (*j*) 13040 ÷ 40   (*k*) 13241 ÷ 41

*11.3 CHANGING BASES

A certain number is written 432 in base five. How is that number expressed in base ten? How can the number we usually write as 87 (in base ten) be expressed in base five? Such translation problems shed still more light on our numeration system.

We begin with the first question, which amounts to translating from base five to base ten. What does the base-five numeral 432 really mean? One brought up using base five would answer this (copying our discussion in Chap. 10) by

Figure 11.5

Base-five numeral	Tally picture	Base-ten numeral
1	│	1
10	ЖЖ	5
100	ЖЖ ЖЖ ЖЖ ЖЖ ЖЖ	25
1000	ЖЖ ЖЖ ЖЖ ЖЖ ЖЖ	125

saying that the meaning of each digit in 432 depends on its position, because each digit is understood to be multiplied by a hidden factor. He would go on to explain that the hidden factors in 432 can be shown explicitly by writing it as $4 \cdot 100 + 3 \cdot 10 + 2 \cdot 1$. Here the hidden factors themselves are expressed in base five, but they can be translated by means of Fig. 11.5.

Thus the base-five numeral 432 represents the same number as the base-ten numeral

$$4 \cdot 25 + 3 \cdot 5 + 2 = 100 + 15 + 2 = 117$$

The key to translating from base five into base ten is evidently to translate the hidden factors, 1, 10, 100, and 1000. These look like the hidden factors we are used to in base ten, but this is only because they are written in base five. If the same numbers were written in base ten, they would be 1, 5, 25, and 125. There is clearly a pattern here, since $5^3 = 125$, $5^2 = 25$, $5^1 = 5$. To continue the pattern, 5^0 is defined to be 1, and indeed b^0 is defined to be 1 for any number b except 0.

We can express a number in base ten if it is given in base five. Can we go the other way? For example, what base-five numeral represents the number which is written as 1771 in base ten? The first step is to write (in base ten) the hidden factors for the base-five digits, namely $5^1, 5^1, 5^2, \ldots$. Multiplied out, these are 1, 5, 25, 125, 625, 3125, etc. Since $3125 > 1771$, the numeral we seek needs no digits with hidden factors as large as 3125, so we must write 1771 as

$$p \cdot 625 + q \cdot 125 + r \cdot 25 + s \cdot 5 + t \cdot 1$$

where p, q, r, s, and t are base-five digits. We determine them by a series of divisions by 625, 125, 25, 5, and 1, according to the following schema:

$$
\begin{aligned}
1771 \div 625 &= 2 + \tfrac{521}{625} \\
521 \div 125 &= 4 + \tfrac{21}{125} \\
21 \div 25 &= 0 + \tfrac{21}{25} \\
21 \div 5 &= 4 + \tfrac{1}{5} \\
1 \div 1 &= 1
\end{aligned}
$$

Reading the integer quotients from the top down, we see that the base-five numeral we are looking for is 24041. The logic of this process emerges when we retrace the steps in reverse order. They tell us, successively,

$$21 = 4 \cdot 5 + 1 \cdot 1$$
$$21 = 0 \cdot 25 + 4 \cdot 5 + 1 \cdot 1$$

$$521 = 4 \cdot 125 + 0 \cdot 25 + 4 \cdot 5 + 1 \cdot 1$$
$$1771 = 2 \cdot 625 + 4 \cdot 125 + 0 \cdot 25 + 4 \cdot 5 + 1 \cdot 1$$

Exercises 11.3

1 (*a*) To express 423 (base five) in base ten, follow these steps:

The 3 stands for $3 \cdot 5^0$ or 3 (base ten).
The 2 stands for $2 \cdot 5^1$ or _____ (base ten).
The 4 stands for $4 \cdot 5^2$ or _____ (base ten).
In total, this is _____ (base ten).

(*b*) To express 2304 (base five) in base ten:

The 4 stands for _____ $\cdot 5^0$ or _____ (base ten).
The 0 stands for $0 \cdot$ _____ or 0 (base ten).
The 3 stands for _____ $\cdot 5^2$ or _____ (base ten).
The 2 stands for $2 \cdot 5^3$ or _____ (base ten).
The total is _____, which is the base-ten numeral for 2304 (base five).

2 These numerals represent numbers in base five. Express each in base ten.
  (*a*) 2   (*b*) 12   (*c*) 101   (*d*) 210   (*e*) 431
  (*f*) 1432   (*g*) 4043   (*h*) 121304

3 (*a*) In base ten 58 may be written $2 \cdot 5^2 + 1 \cdot 5^1 + 3 \cdot 5^0$, so 58 (base ten) can be expressed in base five as _____ .
  (*b*) In base ten 148 may be written $1 \cdot 5^3 + 0 \cdot 5^2 + 4 \cdot 5^1 +$_____ $\cdot 5^0$, so the base-five numeral for 148 (base ten) is _____ .
  (*c*) In base ten 417 may be written _____ $\cdot 5^3 + 1 \cdot$ _____ $+ 3 \cdot 5^1 +$_____ $\cdot 5^0$, so the base-five numeral for 417 (base ten) is _____ .
  (*d*) In base ten 229 may be expressed as _____ $\cdot 5^3 + 4 \cdot 5^2 + 0 \cdot 5^1 +$_____ $\cdot 5^0$, so the base-five numeral for 229 (base ten) is _____ .

4 These numerals represent numbers in base ten. Express each number in base five.
  (*a*) 7   (*b*) 40   (*c*) 75   (*d*) 231
  (*e*) 100   (*f*) 492   (*g*) 1103

5 To count in base three we shall use tally pictures like these
  I for one
  II for two
  ⟋⟋ for three
  ⟋⟋I for four

(*a*) Make base-three tally pictures for five, six, seven, and eight.
(*b*) If nine in base three is pictured as in Fig. 11.6, make base-three tally pictures for ten, eleven, twelve, thirteen, and so on up to twenty-six.
(*c*) The base-three numerals for one, two, three and four are 1, 2, 10, and 11, respectively. Compare these with the tally pictures above, then write the base-three numerals for all higher numbers through twenty-six. The tally pictures from parts (*a*) and (*b*) will help.
(*d*) What is the base-three numeral for twenty-seven?

6 Carry out the indicated computations, given that all numerals in this problem are in base three.
  (*a*) $2 + 2$ (*b*) $12 + 1$ (*c*) $22 + 1$
  (*d*) $201 - 120$ (*e*) $1,121 + 222$ (*f*) $22 \cdot 202$
  (*g*) $101 \cdot 202$ (*h*) $211 \div 102$ (*i*) 21^{11} (exponent, too, is in base three)

7 The numeral 10 represents ten in base ten and five in base five. What does it represent in:
  (*a*) Base three   (*b*) Base seven
  (*c*) Base twelve   (*d*) Generalize

8 What number does the numeral 100 represent in:

Figure 11.6

  (*a*) Base ten   (*b*) Base five   (*c*) Base three
  (*d*) Base seven   (*e*) Base nine   (*f*) Base eleven
  (*g*) Generalize

9 (*a*) Express 3 (base ten) in base seven.
  (*b*) Express 5 (base ten) in base nine.
  (*c*) Express 6 (base ten) in base twelve.
  (*d*) Why are these easy?

10 (*a*) Express 8 (base ten) in base eight.
  (*b*) Express 4 (base ten) in base four.
  (*c*) Express 12 (base ten) in base twelve.
  (*d*) Express 36 (base ten) in base six.
  (*e*) Express 49 (base ten) in base seven.
  (*f*) Express 81 (base ten) in base three.
  (*g*) Express 64 (base ten) in base four.
  (*h*) Express 64 (base ten) in base two.

11 (*a*) We may write 97 (base ten) as $1 \cdot 4^3 + 2 \cdot 4^2 + 0 \cdot 4 + 1 \cdot 4^0$, so the base-four numeral for 97 (base ten) is _____ .
  (*b*) Express 111 (base ten) in base four.
  (*c*) Express 275 (base ten) in base four.
  (*d*) Express 493 (base ten) in base six.
  (*e*) Express 582 (base ten) in base nine.

12 (*a*) Express 31 (base four) in base ten.
  (*b*) Express 57 (base eight) in base ten.
  (*c*) Express 347 (base nine) in base ten.
  (*d*) Express 10101 (base two) in base ten.

13 In any system of numeration with a base greater than ten, new digits are required. In base twelve, let α stand for ten and β for eleven.
  (*a*) What is 10 (base twelve) in base ten?
  (*b*) What is 20 (base twelve) in base ten?
  (*c*) What is 23 (base twelve) in base ten?
  (*d*) What is 1α (base twelve) in base ten?
  (*e*) What is $\alpha 0$ (base twelve) in base ten?
  (*f*) What is $\alpha \beta 2$ (base twelve) in base ten?
  (*g*) What is 35 (base ten) in base twelve?

14 Fractions in other number bases are much like those in base ten. For example, $\frac{31}{143}$ (base five) represents the number which we know as $\frac{16}{48}$ (base ten) or $\frac{1}{3}$, since 31 (base five) = 16 (base ten) and 143 (base five) = 48 (base ten).
  Find base-ten fractions which represent the same numbers as these base-five fractions:

  (*a*) $\frac{3}{12}$ (*b*) $\frac{4}{10}$ (*c*) $\frac{10}{14}$ (*d*) $\frac{32}{100}$ (*e*) $\frac{103}{132}$

15 Which (if any) of the base-five fractions in Prob. 14 are not in lowest terms? Reduce them.

16 Try these computations with base-five fractions.

  (*a*) $\frac{2}{3} + \frac{1}{4}$ (*b*) $\frac{13}{20} - \frac{2}{11}$ (*c*) $\frac{1}{3} \cdot \frac{2}{3}$ (*d*) $\frac{3}{4} \div \frac{4}{12}$

17 Fractions in other number bases can be expressed in a form similar to decimal form. For example, 3.12 (base five) translated into base ten means

$$3 \cdot 1 + 1 \cdot \frac{1}{5} + 2 \cdot \frac{1}{5^2}$$

which is $3\frac{7}{25}$ or 3.28 in decimal form. Express these base-five fractions (they are not decimals; should they be called "quintimals"?) in decimal form.

(a) 2.43   (b) 3.32   (c) 0.0021

18   Would it make any sense at all to use a negative number such as -2 or -10 as the base of a system of numeration? Can you think of any advantages or disadvantages?
(Martin Gardner's April 1973 column in *Scientific American* is about this.)

11.4 FUN WITH BASE TWO

Here is an old but enjoyable trick. Someone thinks of an integer from 1 through 15. It appears in one or more of the columns in Fig. 11.7, and the trick is to find it from yes or no answers to the questions: Is your number in the left-hand column? Is it in the column next to the left? Is it in the column next to the right? Is it in the right-hand column? Of course, with these answers anyone can find the number by searching; the trick is to do it instantly, without looking. Despite tongue-in-cheek references to eyes in the back of the head and photographic memory, people try to figure out the trick. Helping them do so can be very enjoyable.

A hint can be given by modifying the questions somewhat like this: Is your number in the column with the 8? Is it in the column with the 4? In a few minutes, someone usually realizes that you are simply adding whichever of the numbers 8, 4, 2, and 1 head the columns for which the replies were yes to find the person's number. A session with this trick in a classroom often ends up with students playing the trick on each other and copying Fig. 11.7 so they can play the trick on others after school.

Knowing how to play the trick is not the same as knowing why it works, but it helps. The mathematical basis for the trick is in the design of the table in Fig. 11.7. This table has many interesting patterns, but the essential observation is that any integer from 1 to 15 can be expressed as a sum of some or all of 1, 2, 4, and 8. Since these are powers of 2, it is natural to suspect this trick has something to do with base-two numeration, and indeed it does. Can you find the connection? We postpone giving it until we consider a second trick which looks different at first but is actually closely related.

Make a deck of 16 cards like those in Fig. 11.8 out of index cards, spacing

Figure 11.7

8	4	2	1
9	5	3	3
10	6	6	5
11	7	7	7
12	12	10	9
13	13	11	11
14	14	14	13
15	15	15	15

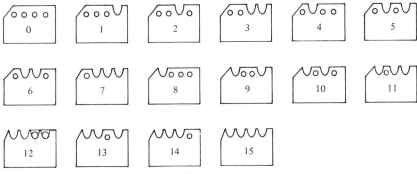

Figure 11.8

the holes and notches carefully so that when the cards are stacked with their cut corners at the upper left, there are four holes through the entire deck.

Shuffle the cards, then stack them with the cut corners at the upper right, so that the cards face away from you and can be read by your audience or class. Now put a toothpick through the leftmost hole in the deck (as you see it) and lift. The cards with holes in the leftmost position will be picked up by the toothpick while those with notches in that position are left behind. Put the cards you picked up on the side of the deck toward the audience (cut corners still at upper left) and remove the toothpick. Repeat the process with the other positions, each time moving one position further to the right (as you see it from behind the cards). When you are done, the cards will be in order. Can you see why this works? How is this related to the first trick?

To see why this second trick works is not hard. Since the 0 card has holes in all four positions and no notches, it is picked up and moved toward the front each time the toothpick is lifted. Any other card has a notch in at least one position, which causes it to stay behind at some stage when the 0 card is picked up. Thus the 0 card is eventually moved ahead of every other card. Likewise the 1 card is passed by the 0 card at the first step, but thereafter it is picked up each time, so it passes all cards with higher numbers, but it never passes the 0 card. One can continue in this manner, but although it explains the trick, it is somehow not as enlightening as might be hoped.

A more interesting question is how the two tricks are related. The first trick depends on the fact that each number corresponds to a particular sequence of yes and no answers, while the second depends on the fact that each numbered card has a particular pattern of holes and notches. Are these patterns related to each other? Consider the 9 card in Fig. 11.9. Reading from left

Figure 11.9

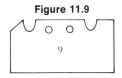

to right, its pattern is notch, hole, hole, notch. Now if you had thought of 9 in
the first trick, you would have replied yes (9 is in the left hand column of Fig.
11.7), no, no (9 is in neither of the middle columns), yes (9 is in the right-hand
column). The correspondence

notch	hole	hole	notch
↕	↕	↕	↕
yes	no	no	yes

is rather striking. Evidently "notch" corresponds to "yes" and "hole" to "no."
This illustrates something rather common in mathematics, a single pattern
turning up in settings which at first seem unrelated. What is the pattern in this
case? A clue is supplied by the binary (base-two) numeral for nine, 1001. Compare it to the nine card. Check a few other cases. What do you notice? Given
the binary numeral 1101 for thirteen, can you sketch the 13 card? Try it, and
use Fig. 11.8 to check yourself. Once you have found the connection to base-
two, you can see why the first trick works and how Fig. 11.7 was made up. For
example, since $13 = 1 \cdot 8 + 1 \cdot 4 + 0 \cdot 2 + 1 \cdot 1$, thirteen appears in Fig. 11.7
beneath 8, 4, and 1 but not beneath 2. The table was made up by heading the
columns with 1, 2, 4, and 8 and then writing each other number under whichever of these were needed. By putting in a new column headed 16, the table in
Fig. 11.7 may be expanded to higher numbers. Then $17 = 16 + 1$ will be written
beneath 16 and 1, $18 = 16 + 2$ will go beneath 16 and 2, and so on.

The base-two pattern turns up in widely scattered parts of mathematics.
As a third example, here is a curious way to multiply integers, sometimes
called *Russian peasant multiplication*, even though it is essentially the method
used in ancient Egypt.

We illustrate with $75 \cdot 231$. At each step we double one factor, beginning
say, with 231, and halve the other, blithely ignoring remainders, and arranging
the work in columns:

75	231
37	462
18	924
9	1848
4	3696
2	7392
1	14784

Next cross out any row in which the left-hand number is even, like this

75	231
37	462
~~18~~	~~924~~
9	1848
~~4~~	~~3696~~
~~2~~	~~7392~~
1	14784

Now add the remaining numbers in the right-hand column to get 17,325 which is indeed $231 \cdot 75$. Try a few for yourself, checking the results by more conventional methods. Can you see why this works? *Hint*: The numbers added in our example are $231 \cdot 1 + 231 \cdot 2 + 231 \cdot 8 + 231 \cdot 64$, and $75 = 1 + 2 + 8 + 64$.

Exercises 11.4

1   In Sec. 5.3 we met the perfect numbers, of which the five smallest are 1, 6, 28, 496, and 8,128. (*a*) Express each of these in base two. (*b*) Look for a pattern in your answers to part (*a*). Can you explain it? *Hint*: See Euclid's formula in Sec. 5.3.

CHAPTER 12
MULTIPLIERS

In this chapter we shall build multipliers much like the adder of Sec. 2.4, so you may want to review that section before beginning this chapter. Our first multipliers will be crude, but later ones will be remarkably practical. Along the way we shall extend and interrelate ideas introduced earlier, particularly those about exponents, decimals, and negative numbers. This chapter is both a culmination of previous work and a glimpse of more advanced ideas.

You will understand this chapter best if you actually build the multipliers yourself as we go along. For accurate spacing, which is important, squared paper will be a big help. The most convenient size has lines $\frac{1}{5}$ inch apart, but $\frac{1}{4}$-inch paper will do.

12.1 A FIRST TRY

We use the same dot pattern (Fig. 12.1) for the multiplier as we did for the adder, but we number the dots differently. We began numbering the dots for the adder with a horizontal row of 0s, because 0 may be added to any number without changing it. Does any number play a corresponding role for multiplication? No number is changed if it is multiplied by 1, and so we begin by numbering one row of dots 1, as in Fig. 12.2. How shall we number dot A? A natural idea is to try 2, so we do. Now if we are to multiply by lining up factors in the outside columns and reading their product in the center (which is analogous to the adder), then B must also be 2, since it is the product of the right-hand 1 and A. Now what about C? When it is multiplied by the 1 on the left, the product is B, which we decided is 2, so $C \cdot 1 = 2$, and $C = 2$. That in turn forces us to put a 4 at D, since $AC = D$ and A and C are both 2. We now have Fig. 12.3. What is E? If the device is to multiply, the dotted line in Fig. 12.3, must represent $E \cdot 1 = 4$, and so $E = 4$. Likewise $G = 4$, and this forces us to make $F = 2 \cdot 4 = 8$ and $H = 4 \cdot 4 = 16$. Fill in your multiplier for a few more steps, and

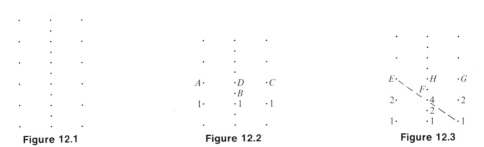

Figure 12.1   Figure 12.2   Figure 12.3

then look it over. What do you notice? What are its advantages and disadvantages? When you have considered these questions, read on.

Your multiplier should look like Fig. 12.4. It is hardly a practical device, since any multiplier worthy of the name should be able to multiply by 3, 5, 6, and 7, to name but a few of the numbers that don't appear. One might hope to put these numbers in, for instance by inserting 3s midway between the 2s and 4s. But this cannot be correct, as then we would read that $3 \cdot 3 = 8$. Evidently the 3 belongs a bit closer to 4 than to 2, but to place it and other numbers like 5, 6, and 7 with reasonable accuracy is quite difficult. Still, for all its problems, the device in Fig. 12.4 is not a total failure, since at least for the numbers we labeled it does indeed multiply. Before we go on to build more multipliers let us consider how this one works. The key is in the numbers labeled. What do you notice about them? *Hint*: If you don't recognize them, factor them into primes.

You no doubt recognized 1, 2, 4, 8, 16, etc., as powers of 2, and in these terms the multiplying device can be rewritten as in Fig. 12.5. Look at this carefully, paying particular attention to the exponents, before you read on.

A careful look shows that Fig. 12.5 is an adder for exponents. Why should this be so? Why, for example does $2^3 \cdot 2^5 = 2^{3+5}$? The answer lies in the meanings of these symbols: $2^3 = 2 \cdot 2 \cdot 2$, and $2^5 = 2 \cdot 2 \cdot 2 \cdot 2 \cdot 2$. Naturally $2^3 \cdot 2^5 = (2 \cdot 2 \cdot 2)(2 \cdot 2 \cdot 2 \cdot 2 \cdot 2)$, which is 2^8 when the parentheses are removed. A similar line of reasoning holds for any two exponents, an observation which plays a key role in mathematics. Comparing Figs. 12.4 and 12.5 shows why it is natural to define 2^0 as 1 and similarly $b^0 = 1$ for any number $b \neq 0$.

	Figure 12.4			Figure 12.5	
32·	·1024	·32	2^5·	2^{10}·	·2^5
	512·			·2^9	
16·	·256	·16	2^4·	2^8·	·2^4
	128·			·2^7	
8·	·64	·8	2^3·	2^6·	·2^3
	32·			·2^5	
4·	·16	·4	2^2·	2^4·	·2^2
	8·			·2^3	
2·	·4	·2	2^1·	2^2·	·2^1
	2·			·2^1	
1·	·1	·1	2^0·	2^0·	·2^0

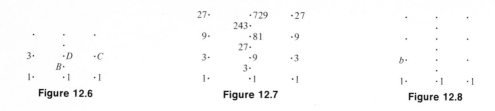

Figure 12.6 **Figure 12.7** **Figure 12.8**

Exercises 12.1

1 The adder could also be used to subtract, and it seems plausible that subtracting exponents in Fig. 12.5 might correspond to division in Fig. 12.4. Figure out how to use the multiplier in Fig. 12.4 to divide.

12.2 MORE AND BETTER MULTIPLIERS

How can our crude multiplier be improved? Can we make one that has 3 on it? In our first try we put 4s where we might have hoped for 3s, but we had to do that to make the multiplier work at all. Once the first 2 was put in, we had no more choice in numbering the other dots, but putting in the 2 was the result of a choice. What if we had chosen 3 instead, as in Fig. 12.6? That would make $B = 3$, and this in turn would determine the value of C. When we continue this way, all the other numbers on the multiplier will be determined. Can you find them? Try before reading on, and then check your work against Fig. 12.7.

Before the multiplier had no 3s, but now it has powers of 3 and nothing else! This one is even less practical than the first, for it will be very hard to place 2, 4, 5, 6, 7, 8 (and most other numbers) on it.

It seems we can choose a single number on our multiplier, and having chosen it all the other numbers will be determined. If we call this number b, as in Fig. 12.8, we shall be led to the multiplier in Fig. 12.9. This is an adder for exponents, but it is practical? That depends entirely on the choice of b. Choosing the value of 2 for b did not lead to very useful results, and we did even worse with $b = 3$. The problem is that powers of 2 increase too fast, and powers of 3 increase even faster. What if we made a multiplier with $b = 4$? Would it be better? What if we tried $b = 1$? Values of $b \geq 2$ are impractical because their powers increase too fast, but powers of 1 do not increase at all. What can we do?

Figure 12.9

$$
\begin{array}{ccc}
b^3 \cdot & \cdot b^6 & \cdot b^3 \\
 & b^5 \cdot & \\
b^2 \cdot & \cdot b^4 & \cdot b^2 \\
 & b^3 \cdot & \\
b \cdot & \cdot b^2 & \cdot b \\
 & b \cdot & \\
1 \cdot & \cdot 1 & \cdot 1
\end{array}
$$

Figure 12.10

$$
\begin{array}{ccc}
1.1^3 \cdot & \cdot 1.1^{16} & \cdot 1.1^3 \\
 & \cdot 1.1^{15} & \\
1.1^2 \cdot & \cdot 1.1^{14} & \cdot 1.1^2 \\
 & \cdot 1.1^{13} & \\
1.1 \cdot & \cdot 1.1^{12} & \cdot 1.1 \\
 & \cdot 1.1 & \\
1 \cdot & \cdot 1 & \cdot 1
\end{array}
$$

The only hope is to try a value of b less than 2 but greater than 1. We shall try to build a better multiplier with $b = 1.1$. In terms of powers of 1.1 this new multiplier is shown in Fig. 12.10. Just how many steps to carry this is still to be determined, but we now begin computing the powers. By direct multiplication we have

$$
\begin{array}{r}
1.1 \\
\times\ 1.1 \\
\hline
1\ 1 \\
1\ 1 \\
\hline
1.2\ 1 = 1.1^2
\end{array}
$$

and

$$
\begin{array}{r}
1.1^3 = (1.1)^2\,(1.1) = 1.21 \\
\times\quad 1.1 \\
\hline
1\ 2\ 1 \\
1\ 2\ 1 \\
\hline
1.3\ 3\ 1 = 1.1^3
\end{array}
$$

Likewise $1.1^4 = (1.1)\,(1.1)^3 = (1.1)\,(1.331)$, and

$$
\begin{array}{r}
1.3\ 3\ 1 \\
\times\quad 1.1 \\
\hline
1\ 3\ 3\ 1 \\
1\ 3\ 3\ 1 \\
\hline
1.4\ 6\ 4\ 1 = 1.1^4
\end{array}
$$

Continuing this way, we compute each power by multiplying the previous power by 1.1. Multiplication by 1.1 is particularly simple; can you find a short-cut for it? It is summed up in the following schema.

$$
\begin{array}{r}
1.1 = 1.1^1 \\
+\ 1.1 \\
\hline
1.2\ 1 = 1.1^2 \\
+\ 1.2\ 1 \\
\hline
1.3\ 3\ 1 = 1.1^3 \\
+\ 1.3\ 3\ 1 \\
\hline
1.4\ 6\ 4\ 1 = 1.1^4 \\
+\ 1.4\ 6\ 4\ 1 \\
\hline
1.6\ 1\ 0\ 5\ 1 = 1.1^5 \\
+\ 1.6\ 1\ 0\ 5\ 1 \\
\hline
1.7\ 7\ 1\ 5\ 6\ 1 = 1.1^6 \\
+\ 1.7\ 7\ 1\ 5\ 6\ 1 \\
\hline
1.9\ 4\ 8\ 7\ 1\ 7\ 1 = 1.1^7 \\
+\ 1.9\ 4\ 8\ 7\ 1\ 7\ 1 \\
\hline
2.1\ 4\ 3\ 5\ 8\ 8\ 8\ 1 = 1.1^8
\end{array}
$$

n	Approximate value of 1.1^n	n	Approximate value of 1.1^n	n	Approximate value of 1.1^n
1	1.1	18	5.5585	35	28.0935
2	1.21	19	6.1143	36	30.9028
3	1.331	20	6.7257	37	33.9930
4	1.4641	21	7.3982	38	37.3923
5	1.6105	22	8.1380	39	41.1315
6	1.7715	23	8.9518	40	45.2446
7	1.9486	24	9.8469	41	49.7690
8	2.1434	25	10.8315	42	54.7459
9	2.3577	26	11.9146	43	60.2204
10	2.5934	27	13.1060	44	66.2424
11	2.8527	28	14.4166	45	72.8666
12	3.1379	29	15.8582	46	80.1532
13	3.4516	30	17.4440	47	88.1685
14	3.7967	31	19.1884	48	96.9853
15	4.1763	32	21.1072	49	106.6838
16	4.5939	33	23.2179	50	117.3521
17	5.0532	34	25.5396		

Figure 12.11

This process is direct and simple, but each power of 1.1 is a digit longer than the one before, which makes both more work and more chance for error (a particularly insidious possibility here, as any mistake in the computation of one power will be carried along in the computation of higher powers). Therefore we throw away all digits after the first four decimal places in our computations. This will introduce some error, but our multiplier will still be surprisingly accurate. The resulting approximate powers are tabulated in Fig. 12.11, up to 1.1^{50}, which is approximately 117.3521. We shall see that this is high enough to build a very useful multiplier.

Rounding these figures off to the nearest hundredth yields the multiplier in Fig. 12.12.

Exercises 12.2

1 Use the multiplier in Fig. 12.12 to find:
   (a) 2.36×8.14 (b) 8.95×1.95 (c) 6.11×4.18
2 In building the multiplier we used two-decimal-place approximations to powers of 1.1 instead of the powers themselves, sacrificing accuracy for convenience. This problem will give you some idea of the seriousness of the error introduced.
   (a) Compute the three products from Prob. 1 by hand, and compare your results with those found with the multiplier. Yours (if you multiply correctly) will be the true products, while those from the multiplier are approximations.
   (b) In each case find which of the true answer and the approximation is smaller and subtract it from the other. This is the error incurred by using the multiplier.
   (c) The importance of an error depends not only on its size but on the size of the quantity in question. (An error of a millimeter can be important in surgery, but an error of a foot or more may be unimportant in surveying, where the distances involved are much larger.) In each of the three cases compare the error with $\frac{1}{100}$ times the true product. What do you conclude about the accuracy of the multiplier?
3 Find out how to use the multiplier to square numbers.
4 As it stands the multiplier can multiply only factors between 1 and 10.83. In Sec. 12.4 we shall

discuss how to use it to multiply numbers outside this range, but you may be able to figure it out for yourself now. Do so if you can.
5 How could you build an even more accurate multiplier? What practical problems might you encounter? (This is discussed in Sec. 12.5.)

12.3 NEGATIVE EXPONENTS

The multipliers are simply adders for exponents. The adder works for negative exponents as well as positive, but we have not extended the multiplier to nega-

Figure 12.12

10.83	117.35	10.83	*NOTE:* For clarity
	106.68		the decimal points
9.85	96.99	9.85	here serve as
	88.17		the dots in the
8.95	80.15	8.95	basic multiplier
	72.87		pattern.
8.14	66.24	8.14	
	60.22		
7.40	54.75	7.40	
	49.77		
6.73	45.24	6.73	
	41.13		
6.11	37.39	6.11	
	33.99		
5.56	30.90	5.56	
	28.09		
5.05	25.54	5.05	
	23.22		
4.59	21.11	4.59	
	19.19		
4.18	17.44	4.18	
	15.86		
3.80	14.42	3.80	
	13.11		
3.45	11.92	3.45	
	10.83		
3.14	9.85	3.14	
	8.95		
2.85	8.14	2.85	
	7.40		
2.59	6.73	2.59	
	6.11		
2.36	5.56	2.36	
	5.05		
2.14	4.54	2.14	
	4.18		
1.95	3.80	1.95	
	3.45		
1.77	3.14	1.77	
	2.85		
1.61	2.59	1.61	
	2.36		
1.46	2.14	1.46	
	1.95		
1.33	1.77	1.33	
	1.61		
1.21	1.46	1.21	
	1.33		
1.1	1.21	1.1	
	1.1		
1.	1.	1.	

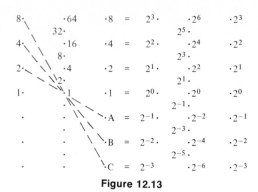

Figure 12.13

tive exponents because we have not defined them. What does something like 2^{-3} mean? How can negative exponents be given a sensible meaning?

The analogy between multiplication and addition is helpful here. We first met multiplication as shorthand for repeated addition, but later we generalized the idea of multiplication to negative integers and fractions. Exponents were first introduced as shorthand for repeated multiplication. How can this be generalized?

For now we shall confine our attention to defining negative integer exponents, using clues from the adder and multiplier.

If a suitable generalization can be found for the multipliers in Fig. 12.13, it will have to follow the patterns there, so $A = 2^{-1}$, $B = 2^{-2}$, and $C = 2^{-3}$. What numbers are these in more familiar form?

The dotted lines in Fig. 12.13 show that $2A = 1$, $4B = 1$, and $8C = 1$, so we must have $A = \frac{1}{2}$, $B = \frac{1}{4}$, and $C = \frac{1}{8}$. Therefore we define 2^{-1} to be $\frac{1}{2}$, 2^{-2} to be $\frac{1}{4}$, and $2^{-4} = \frac{1}{16}$, $2^{-5} = \frac{1}{32}$, and in general $2^{-n} = 1/2^n$.

This discussion has been based on powers of 2, but the reasoning applies as well to powers of other numbers. Extending the multiplier in Fig. 12.9 below 1 (which is b^0) leads to defining $b^{-1} = 1/b$, $b^{-2} = 1/b^2$, and in general $b^{-n} = 1/b^n$ for any number b except 0. This definition may not seem natural at first, but it fits all the patterns. All the multipliers are based on powers of some number b other than 0, and all are adders for exponents. For positive exponents, the key rule is that if n factors of b are followed by m factors of the same number b, there are $n + m$ factors of b in all. That is,

$$b^n b^m = \underbrace{(b \cdot b \cdots b)}_{n \text{ factors}} \underbrace{(b \cdot b \cdots b)}_{m \text{ factors}} = \underbrace{b \cdot b \cdots b}_{n + m \text{ factors}} = b^{n+m}$$

Defining $b^{-n} = 1/b^n$ extends the rule $b^n \cdot b^m = b^{n+m}$ to negative values of m and n so that the multipliers are adders for negative exponents as well as positive.

You have already met one application of this idea. By our definition,

$$10^{-1} = \frac{1}{10}, \qquad 10^{-2} = \frac{1}{10^2} = \frac{1}{100}, \qquad \text{etc.,}$$

so a decimal like 25.37 means $2 \times 10^1 + 5 \times 10^0 + 3 \times 10^{-1} + 7 \times 10^{-2}$. Thus the hidden factors in decimal notation are all powers of 10. Similarly, the hidden factors in base five are all powers of five, and so on.

Exercises 12.3

1 Why was b^{-n} not defined for $b = 0$?

2 Find:

(a) 3^{-2} (b) 10^{-4} (c) 5^{-3} (d) 2^{-4}

(e) 7^{-1} (f) $\left(\frac{1}{2}\right)^{-1}$ (g) $\left(\frac{2}{3}\right)^{-2}$ (h) $\left(-\frac{2}{3}\right)^{-2}$

(i) $\left(\frac{3}{5}\right)^{-2}$ (j) $\left(-\frac{3}{4}\right)^{-3}$ (k) $(0.03)^{-1}$

3 Extend the multiplier in Fig. 12.9 to negative powers of b.

4 Use the rule $a^n \times a^m = a^{n+m}$ to compute these:

(a) $10^{-2} \times 10^5$ (b) $8^{-3} \cdot 8^4$ (c) $2^7 \cdot 2^{-4}$ (d) $3^5 \cdot 3^{-6}$

(e) $5^{-17} \cdot 5^{14}$ (f) $10^{-21} \times 10^{50}$ (g) $10^{-6} \times 10^7$

5 Complete:

(a) Division is the inverse of _____.

(b) Subtraction is the inverse of _____.

(c) To multiply powers of the same number _____ their exponents.

(d) Dividing powers of the same number involves _____ their exponents.

(e) Specifically, $a^n \div a^m = a^{\boxed{}}$ (fill in box.)

6 Use the rule from Prob. 5 to find:

(a) $10^4 \div 10^7$ (b) $10^{-3} \div 10^{-5}$ (c) $10^{-8} \div 10^{-6}$

(d) $10^{13} \div 10^{18}$ (e) $10^{-5} \div 10^4$

7 Many people learn to place the decimal point in a product by the following rule. Put as many digits to the right of the decimal point in the product as there are in all the factors together. Surprisingly few people understand why this works, but this problem will show it.

(a) Express these decimals as "ordinary" fractions with powers of 10 in the denominators:

(1) 0.47   (2) 5.31   (3) 23.0007   (4) 10.3

(b) In part (a), how does the exponent in the denominator compare with the number of digits to the right of the decimal point?

(c) To multiply two decimals (we illustrate with 2.03 and 3.174):

    1 Rewrite the problem with "ordinary" fractions

$$\frac{203}{100} \times \frac{3{,}174}{1{,}000}$$

    2 Express the denominators with powers of 10:

$$\frac{203}{10^2} \cdot \frac{3{,}174}{10^3}$$

    3 Multiply the fractions, adding exponents in the denominator:

$$\frac{203 \times 3{,}174}{10^2 \times 10^3} = \frac{641{,}322}{10^5}$$

4 Restate the product in decimal form. How many digits are to the right of the decimal point?

(*d*) Follow the steps in part (*c*) with 2.114 × 3.0027.

12.4 SCIENTIFIC NOTATION

All the numbers along the sides of the multiplier in Fig. 12.12 are between 1 and 10.83, but we can nonetheless use it to multiply numbers outside this range. For example, to multiply 5,050 by 133, we write 5,050 as 5.05×10^3 and 1.33×10^2. From the multiplier, $5.05 \times 1.33 = 6.73$, so $(5.05 \times 10^3)(1.33 \times 10^2) = 6.73 \times 10^{3+2} = 673,000$. The idea is to express each factor as a number between 1 and 10 times a power of 10. Then multiply the numbers between 1 and 10 on the multiplier and take care of the powers of 10 separately.

The slide rule is related to the multiplier. People who often use a slide rule find it convenient to express most numbers in the form used above, as the product of a power of 10 and a number between 1 and 10. This is known as *scientific notation* because scientists use it so much. The numbers in scientific work are sometimes so very large or small that they are clumsy in ordinary decimal notation. A basic constant in chemistry, for example, is *Avogadro's number*, which is about

$$602,000,000,000,000,000,000,000$$

One must count the 0s to read or write this, but in scientific notation there is no problem: it is simply 6.02×10^{23}.

Exercises 12.4

1 Express these numbers in scientific notation:
(*a*) 400,000,000,000 (somewhere near the number of dollars in our national debt)
(*b*) 16090000 (approximately the number of centimeters in a mile)
(*c*) 299,792,500 (the speed of light in meters per second)
(*d*) 0.00003125 (the number of tons in an ounce)
(*e*) 0.000000001 (the number of cubic meters in a cubic millimeter)

2 Use the multiplier in Fig. 12.12 to find these products and quotients. Leave your answers in scientific notation.
(*a*) $(6.11 \times 10^4)(3.45)$ (*b*) $(2.85 \times 10^2)(4.59 \times 10^5)$
(*c*) $(1.77 \times 10^9)(7.4 \times 10^{-3})$ (*d*) $(6.73 \times 10^{-2})(4.59 \times 10^{-7})$
(*e*) $(11.92 \times 10^2) \div (3.80 \times 10^2)$ (*f*) $(23.22 \times 10^3) \div (9.85 \times 10^3)$
(*g*) $(54.75 \times 10^{-1}) \div (6.11 \times 10^4)$ (*h*) $(45.24 \times 10^{10}) \div (6.73 \times 10^{-5})$
(*i*) $(3.14 \times 10^{-2}) \div (2.59 \times 10^{-1})$

3 Use the multiplier in Fig. 12.12 to find:
(*a*) 25.9×3.45 (*b*) $2,590 \times 3.45$ (*c*) 4.18×0.0459
(*d*) $4,180 \times 4,590$ (*e*) $814,000 \times 0.214$ (*f*) 0.161×0.00000556
(*g*) $25.54 \div 0.0673$ (*h*) $0.002111 \div 0.00038$

12.5 SOME BACKGROUND

Our multiplier is comparable in accuracy to a crude slide rule, but this is not accurate enough for the most careful work. The methods discussed in this chapter were first developed between 1610 and 1620 by Jobst Bürgi, a Swiss watchmaker, and independently by John Napier[1] in Scotland, but even in those days seven-figure accuracy was needed for scientific calculations.

To build a more accurate multiplier is simple in principle: it must merely consist of powers of a number even smaller than 1.1 (though still larger than 1). For seven-figure accuracy such a multiplier would have to be made with powers of 1.0000001. Bürgi actually made a table like this for 1.0001, a huge job, since it takes over 23,000 steps to reach 10, and an error in any step would have ruined the table from there on. Napier began a related project, equivalent in difficulty to making a table based on powers of 1.0000001, but the enormous size of the job (it would take over 23 million steps to reach 10) soon turned him to a more theoretical approach which led to his invention of *logarithms*.

Of course a multiplier with 23,000 steps would be impractical even if it could be built without error. It would have to be printed on such a long scroll of paper that lining up dots accurately would be difficult. Consequently the early pioneers did not build multipliers like ours but worked directly from tables like that in Figure 12.11. All calculations that can be done with the multiplier in Fig. 12.12 can be done directly from the table in Fig. 12.11. For example, to multiply 2.36 by 2.85, we find in the table that 2.36 is approximately 1.1^9 and 2.85 is approximately 1.1^{11}. Therefore, approximately,

$$2.36 \times 2.85 = 1.1^9 \times 1.1^{11} = 1.1^{9+11} = 1.1^{20}$$

The table shows this to be about 6.7257 or, rounded off to the nearest hundredth, 6.73, which could have been read from the multiplier. Tables of logarithms are based on this idea and are widely available to seven- and even ten-place accuracy.

Today the rapid development of computers has made computation with logarithms obsolete, but the underlying theory of exponents remains fundamental in algebra and calculus, both of which have countless applications.

Exercises 12.5

1 Suppose you built a multiplier based on powers of 1.0000001 on a scroll of paper and you spaced the (roughly) 23,000 numbers in each outer column at $\frac{1}{4}$-inch intervals. (This is quite close, as it leaves dots only $\frac{1}{8}$ inch apart in the center column.) About how long would the scroll of paper be?

[1]John Napier (1550–1617) of Merchiston, Scotland, was best known in his own lifetime for a popular and vehemently anti-Catholic book, in which he tried to prove that the Pope was Antichrist and that the world would end in 1786. For him mathematics and science were recreations, but his mathematics will far outlive his theology. Gridgeman's article (see p. 224, Bibliography) is a fascinating account of Napier's life and work.

2 Carry out these computations by means of the table in Fig. 12.11, using Fig. 12.12 only for a
   rough check.
   (a) 8.1380×2.3577 (b) $555.85 \times 3,451.6$ (c) 6.7257^2
   (d) $54.7459 \div 6.1143$ *(e) 21.434^3 *(f) 1.9486^6

CHAPTER 13
AREA

From here on most of our work will have a geometric flavor. As before, you will find that you are able to discover a great deal of mathematics for yourself.

13.1 AREA

We begin with an exploratory exercise designed to develop your geometric intuition and at the same time supply data for later work. Intuitively, area is a measure of how much ground something covers. We shall use a square like the one at the upper left in Fig. 13.1 as our basic unit of area. Try to find the areas of the shapes in Fig. 13.1, expressing them in terms of the square unit. Do not use formulas you may have learned before, but instead try to visualize each problem. The first ones are straightforward, but some of the later ones may challenge you. If so, you may find it helpful to use your knowledge from the previous cases. For example, here are three ways one class found the area of triangle E:

1. Fit it together with A and D to form a rectangle.
2. Put it next to triangle A to make triangle D.
3. View triangle E as half of shape B.

13.2 PARALLELOGRAMS AND TRIANGLES

A four-sided figure with two pairs of parallel sides is called a *parallelogram*. In Fig. 13.1, B and F are parallelograms, but the other shapes there are not. Squares and other rectangles are also examples of parallelograms. If we think of a parallelogram as resting on level ground, then the side it rests on is called the *base* of the parallelogram, and the distance from the top of the parallelo-

Figure 13.1

gram to the ground is called its *height*. Naturally, all this depends on how the parallelogram is viewed. From a different direction we might consider a different side as the base and measure the height from *that* side. There is a simple but important relation between the base and height of a parallelogram and its area. You can find this relationship yourself by recording the base, height, and area of several parallelograms in a table like this and then searching for a

	Fig. 13.1		Fig. 13.2				
Parallelogram	B	F	L	M	N	O	P
Length of base							
Height							
Area							

pattern in the table. Figures 13.1 and 13.2 have some parallelograms whose areas you can find by the methods used before.

When you have found the pattern, consider how you might verify it in general. If the parallelogram happens to be a rectangle, your pattern should agree with our earlier results about such areas (from Sec. 1.2). Can you generalize to parallelograms which are not rectangles?

One way to see this is to imagine cutting off a piece of the parallelogram, as in Fig. 13.3, and moving it to form a rectangle with the same base and height as the original parallelogram.

You may be surprised that the rule for the area of a parallelogram is not summed up here in a formula as it is in most books. The omission is deliberate.

Figure 13.2

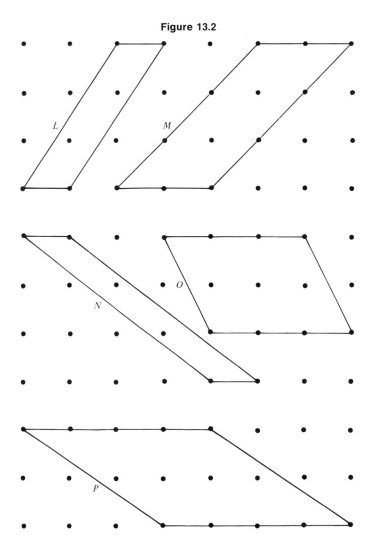

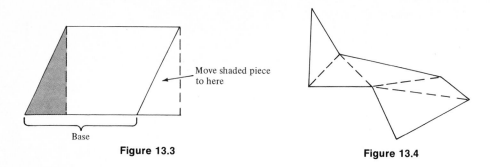

Figure 13.3   **Figure 13.4**

A formula can save the user the trouble of thinking through for himself the rule it embodies. This is often an advantage, but it allows one to use a rule without really understanding it. A main goal of this book is to help you think through the material rather than just accept it on faith, so the formula is not presented here.

One reason for dealing with parallelograms is that they are a preliminary step toward a rule for the areas of triangles. It is especially useful to be able to find the areas of triangles, because areas of more complicated shapes may be found by cutting them up into triangular pieces, as in Fig. 13.4.

The area of a triangle is related to its base and height. Here the *base* is a side of the triangle we view it as resting on, and the height is the distance of the top of the triangle "above ground level," as shown in Fig. 13.5.

To find the relation of the area of a triangle to its base and height, find the base and height for triangles *A*, *D*, *E*, *G*, *H*, *I*, and *K* from Fig. 13.1, and tabulate them, along with their areas. You will find it most convenient to turn

	Triangle						
	A	*D*	*E*	*G*	*H*	*I*	*K*
Length of base							
Height							
Area							

triangles *E* and *H* upside down and *I* on its side. Triangle *J* has been excluded because its base and height are awkward to measure. When you find the relation between the base and height of a triangle and its area, you will probably notice its striking resemblance to the corresponding pattern for parallelograms.

Figure 13.5

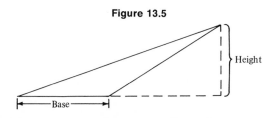

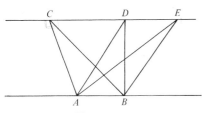

Figure 13.6

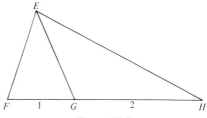

Figure 13.7

Can you explain this? *Hint*: How can two copies of any given triangle be put together to form a parallelogram?

Exercises 13.2

1 Line *AB* is parallel to *CE* in Fig. 13.6.
   (*a*) How do the areas of triangles *ABC*, *ABD*, and *ABE* compare?
   (*b*) How do the areas of triangles *ACE* and *BCE* compare?
2 If *G* in Fig. 13.7 is twice as far from *H* as from *F*, how do the areas of triangles *FEG* and *HEG* compare?
3 *X*, *Y*, and *Z* in Fig. 13.8 are the midpoints of line segments *UY*, *SZ*, and *TX*, respectively. What fraction of the area of triangle *SUT* lies within triangle *YXZ*? (*Hint*: Draw lines *XS*, *YT*, *ZU*.)
4 In 1899 G. Pick found a curious relation concerning shapes which, like those in Fig. 13.1, have straight edges running between dots in any array of *lattice points*. Complete this table and find Pick's rule relating the area of each shape to the number of dots inside it and the number of dots on its boundary. *Hint*: A dot on the boundary may be thought of as halfway inside the shape.

	Shape from Fig. 13.1										
	A	*B*	*C*	*D*	*E*	*F*	*G*	*H*	*I*	*J*	*K*
Number of lattice points inside											
Number of lattice points on boundary											
Area											

Figure 13.8

5 In Fig. 13.9 find:
   (*a*) The area of the outer square.
   (*b*) The area of each of the four triangles.
   (*c*) The area of the tilted inner square. (The dots in each row and column are 1 unit apart.)

13.3 CIRCLES

What is the distance around a circle? How much area does a circle enclose?
These questions and others like them have been asked since ancient times, but
only in recent centuries have they been fully answered. Here we consider some
experiments which yield approximate answers without involving advanced
mathematics.

EXPERIMENT 1

There is a rule relating the distance around a circle to the distance across it.
These distances are called, respectively, the *circumference* and *diameter* of
the circle (Fig. 13.10). Half the diameter is called the *radius*. To find the rule,
assemble several circular objects such as dishes, pans, lampshades, tires, cans,
and hoops. Measure each circumference and diameter as accurately as possible
with a tape measure, and in each case divide circumference by the diameter,
expressing the result in decimal form. What do you observe about your results?

You probably found that in each case the circumference is a bit more than
3 times the diameter. In fact the number by which the diameter of a circle must
be multiplied to get its circumference is independent of the size of the circle.
The precise nature of this number, however, was a major problem for many

Figure 13.9

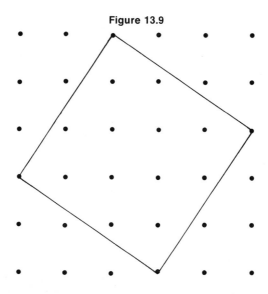

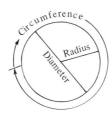

Figure 13.10

centuries. This elusive number was given a special name, π, which is a lower-case Greek letter p. It is written "pi" and pronounced "pie" in English. Of the many estimates of the value of π made by ancient mathematicians, one of the most ingenious was a proof by Archimedes that π lies between $3\frac{10}{71}$ and $3\frac{10}{70}$. Popular confusion about π lasted surprisingly long, however, and as late as 1897 the Indiana State Legislature passed a bill (enthusiastically backed by the State Superintendent of Public Instruction) that would have set the legal value of π at 3 by legislation. Today computers have calculated π to thousands of decimal places, but for elementary work 3.14 and $3\frac{1}{7}$ are usually adequate approximations, even though neither is precisely π.

EXPERIMENT 2: THE AREA OF A CIRCLE AND OTHER AREAS

This experiment requires a good balance scale and very careful work.

1   Carefully cut out the shape whose area is to be found and weigh it.
2   Cut out a square unit of area, drawn on the same kind of paper as the shape whose area is sought and weight this unit square.
3   Divide the weight of the shape in question by the weight of the square unit to estimate the area sought.

If this experiment is done carefully, it can yield surprisingly accurate results. Archimedes used it to guess a formula, which he later proved, for the area of a segment of a parabola.[1] One advantage of weighing is that it may be used even on highly irregular areas, and it can be used to estimate volumes as well as areas.

Exercises 13.3

1   Here are some of the numbers which various scholars of the past have either thought to be π or to closely approximate it. Change each to decimal form to find out how close it is to π, the true value of which (to 30 decimal places) is

[1]Archimedes (287–212 B.C.) is best known for *Archimedes' principle* (a floating body displaces its weight of fluid), but his work ranged far beyond that. By himself he developed parts of mathematics to a level not regained for over 1,800 years. One can only mourn the folly of the Roman soldier who slew him (as he contemplated a mathematical diagram, it is said) when the Romans conquered his home town, Syracuse.

3.14159265358979323846264338327 9

(a) $\left(\frac{16}{9}\right)^2$ (Ahmes the Scribe, Egypt, around 1650 B.C.)

(b) $3\frac{10}{71} < \pi < 3\frac{10}{70}$ (Archimedes)

(c) $\frac{142}{45}$ (Wang Fan, China, around A.D. 250)

(d) $\frac{62.8}{20,000}$ (Aryabhata, India, around A.D. 475)

(e) $\frac{355}{113}$ (Tsu Ch'ung Shih, China, A.D. 430–501)

2 (a) If the earth is a sphere whose diameter is 8,000 miles, how long is the equator? (Ignore mountains on the equator.)

(b) If a string 100 feet longer than the equator were stretched around the earth at a constant height above the equator, how much clearance would there be between the earth's surface and the string? *Hint*: How long is the circle formed by the string? Compare the radius of that circle with the radius of the earth.

*3 If an automobile tire is a perfect circle and its diameter is 2 feet, how many times will it turn if it lasts 40,000 miles? Ignore the fact that it shrinks with wear. (Each time the tire makes one complete rotation, if it does not slip, it rolls a distance equal to its circumference.)

CHAPTER 14
THE PYTHAGOREAN THEOREM

This chapter is about one of the most important relationships in all of mathematics. You probably have at least a passing acquaintance with it, but here we study it in some detail. We begin with a three-part exercise which helps you discover the relationship for yourself and see why it is true. Then we go on to apply it in some simple cases (saving more difficult applications for Chap. 15), and, finally, we consider an attempted generalization which is perhaps the most famous unsolved problem in mathematics today.

14.1 A DISCOVERY EXERCISE

Part I   On squared paper whose lines are a $\frac{1}{4}$ or $\frac{1}{5}$ inch apart place coordinate axes as in Fig. 14.1. Then use a straightedge to draw in the triangles whose vertices are tabulated.

A $(-13, 14), (-13, 19), (-1, 19)$
B $(-6, 3), (0, 3), (0, 11)$
C $(3, 1), (3, 16), (11, 1)$
D $(7, 16), (11, 16), (11, 13)$
E $(-12, 0), (12, 0), (-12, -7)$
F $(12, -2), (-12, -12), (12, -12)$

Figure 14.1

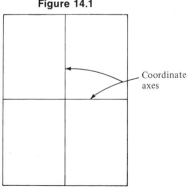

Coordinate axes

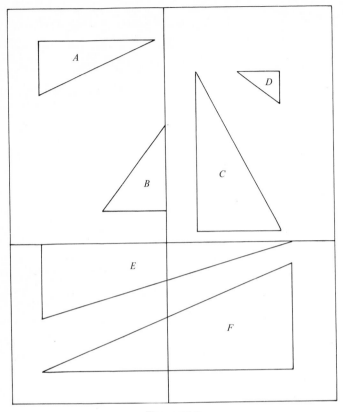

Figure 14.2

These triangles must be plotted accurately to make the discovery easy. Figure 14.2 is a rough sketch of how it should look. How are the lengths of the sides of these triangles related? As a first step, measure all three sides of each triangle. Use as unit of measure the distance between adjacent parallel lines on your paper $\left(\frac{1}{4}\right.$ inch or $\frac{1}{5}$ inch, depending on your paper$\left.\right)$. That way, for example,

Figure 14.3

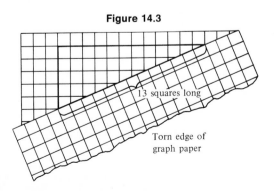

Triangle	Length of shortest side	Length of next to shortest side	Length of longest side
A			
B			
C			
D			
E			
F			

Figure 14.4

the shortest side of triangle A is five boxes long, regardless of the size of the squares on your paper. Two sides of each triangle can be measured simply by counting boxes, but the third side is more awkward. You could use a ruler for this, but because of inaccurate printing the scales of ruler and paper may not be precisely the same. A better way is to improvise a ruler from an edge of the same kind of graph paper, as shown in Fig. 14.3. If your coordinate axes are well centered, you can probably tear off an edge of your paper for this. Tabulate your results as in Fig. 14.4.

The lengths of the three sides of each triangle measured are related by an elegant rule. You may enjoy trying to guess the rule, but do not be disappointed if you cannot find it easily. It is rather subtle, but it will emerge clearly as we go along.

Part II Now compute the lengths of the longest sides of our triangles in part I using the following indirect approach. This will, incidentally, give you a way to check your measurements from part I.

Figure 14.5 shows four triangles just like triangle A arranged to form a square with another tilted square inside it. The outer square is 17 units on a side, so its area is $17^2 = 289$ square units. The four copies of triangle A have a total of 120 square units. (Do you see why? Those in two opposite corners can be paired to make a 5 by 12 rectangle.) So the area of the tilted inner

Figure 14.5

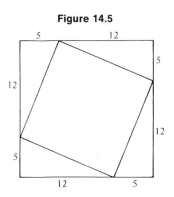

square is $289 - 120 = 169$ square units. Since $13^2 = 169$, the inner square is 13 units long on each edge. But each edge of this inner square is the longest side of a copy of triangle A. In your measurements from part I did you find that the longest side of triangle A is 13 units long?

Use the same reasoning and the accompanying figures to fill in these blanks and check your work on the other triangles from part I.

In Fig. 14.6 length of edge of outer square _____.
Area of outer square _____.
Total area of the four triangles _____.
Area of tilted inner square _____.
Length of edge of inner square _____. Is this the length you measured for the longest side of triangle B?

In Fig. 14.7 length of edge of outer square _____.
Area of outer square _____.
Total area of the four triangles _____.
Area of the tilted inner square _____.
Length of edge of inner square _____. Does this agree with the corresponding measurement (the length of the longest side of triangle C) from part I?

In Fig. 14.8 length of edge of outer square _____.
Area of outer square _____.
Total area of the four triangles _____.
Area of the tilted inner square _____.
Length of edge of inner square _____. Does this agree with the corresponding measurement (the length of the longest side of triangle D) from part I?

Figure 14.6   Triangle *B*.

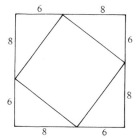

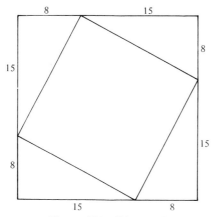

Figure 14.7 Triangle C.

Figure 14.8 Triangle D.

In Fig. 14.9 length of outer square _____.
Area of outer square _____ .
Total area of the four triangles _____ .
Area of the tilted inner square _____.
Length of sides of inner square _____. Does this agree with your measurement for triangle E in part I?

Figure 14.9 Triangle E.

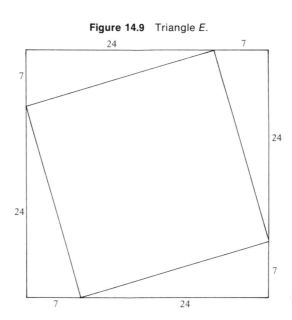

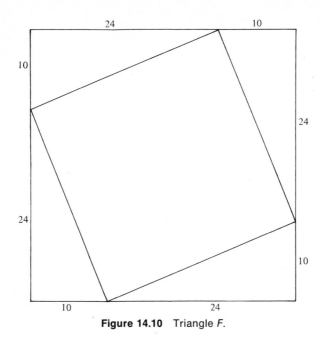

Figure 14.10 Triangle *F*.

In Fig. 14.10 length of outer square _____ .
Area of outer square _____ .
Total area of the four triangles _____ .
Area of tilted inner square _____ .
Length of side of inner square _____ . Does this agree with your earlier measurement for triangle *F* in part I?

Part III To relate the work from part II (where we indirectly measured the longest side of each triangle from part I) to the shorter sides of the triangles, we

Figure 14.11   **Figure 14.12** Triangle *B*.

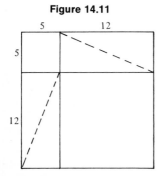

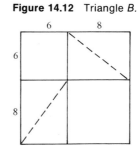

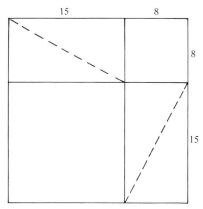

Figure 14.13   Triangle *C*.

Figure 14.14   Triangle *D*.

need only divide up the squares from part II in another way. The square for triangle *A* was 17 units long because 17 is the sum of the lengths of the two shorter sides, 5 and 12, for triangle *A*. Now we divide the same 17 by 17 square as in Fig. 14.11 and compare it with Fig. 14.5. The dotted lines in Fig. 14.11 show that the two 5 by 12 rectangles there have the same total area as the four triangles in Fig. 14.5. Therefore the 12 by 12 square and the 5 by 5 square in Fig. 14.11 have together as much area as the tilted inner square in Fig. 14.5. The numbers bear this out, as $144 + 25 = 169$, or, more concisely, $12^2 + 5^2 = 13^2$. Fill in the blanks below to carry out similar comparisons in the other cases.

In Fig. 14.12 area of 6 by 6 square   _____.
Area of 8 by 8 square   _____.
Add these areas   _____.
Area of tilted inner square in Fig. 14.6 _____.

In Fig. 14.13 area of 8 by 8 square   _____.
Area of 15 by 15 square   _____.
Add these two areas   _____.
Area of tilted inner square in Fig. 14.7 _____.

In Fig. 14.14 area of 3 by 3 square   _____.
Area of 4 by 4 square   _____.
Sum of these two areas   _____.
Area of tilted inner square in Fig. 14.8 _____.

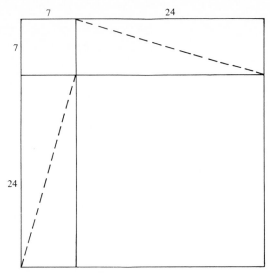

Figure 14.15   Triangle *E*.

In Fig. 14.15 area of 7 by 7 square   _____.
Area of 24 by 24 square   _____.
Sum of these two areas   _____.
Area of tilted inner square in Fig. 14.9 _____.

Figure 14.16   Triangle *F*.

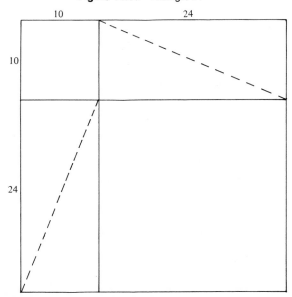

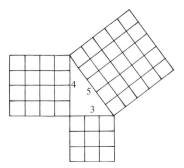

Figure 14.17

In Fig. 14.16 area of 10 by 10 square _____.
Area of 24 by 24 square   _____.
Sum of these areas   _____.
Area of tilted square in Fig. 14.10   _____.

You no doubt found that in each case the areas of the two squares added in part III totaled the area of the tilted square in part II. This means that if you make squares on the two shorter sides of one of the triangles in our discovery exercise, their areas will add up to the area of the square on the longest side. Figure 14.17 shows this for triangle *D*, and similar sketches can be made for the other triangles we worked with. Numerically, this means if all entries in Fig. 14.4 are squared, the first two numbers in each row will add up to the third. This is shown in Fig. 14.18.

Does this pattern hold for other triangles? Yes, provided that (like those we have worked with) they have a square corner. Such a triangle is called a *right triangle*, because the angle associated with a square corner is called a *right angle*. In general, if squares are built on the three sides of a right triangle, the areas of the two smaller squares add up to the area of the largest one. Specific examples of this rule were known in ancient Egypt and many other ancient civilizations, but Pythagoras is considered to have been the

Figure 14.18

Triangle	Square of length of shortest side	Square of length of intermediate side	Square of length of longest side
A	25	144	169
B	36	64	100
C	64	225	289
D	9	16	25
E	49	576	625
F	100	576	676

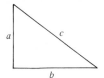

Figure 14.19

first to state it in general terms, and today it is known as the *Pythagorean theorem.*[1]

To verify the Pythagorean theorem in general, we proceed just as we did in the previous cases. Suppose we have a right triangle. Let c stand for the length of the longest side, and let a and b stand for the lengths of the other two sides, as in Fig. 14.19. How are the lengths of the sides related? We shall show that $c^2 = a^2 + b^2$. To see this, consider a square, each side of which is $a + b$ units long. If we divide this square as in Fig. 14.20 and compute the areas of the parts as if this were a box puzzle like those in Sec. 1.4, we find the area is $a^2 + b^2 + 2ab$.

On the other hand, if we divide the same square as in Fig. 14.21, we see it has copies of the original triangle in the corners, so that its area is that of the four triangles plus that of the tilted inner square. The four triangles each have areas $ab/2$, so their total area is $4\,ab/2 = 2ab$. (You can also see this by putting the four triangles together to form two rectangles like those in Fig. 14.20.) The tilted inner square has area c^2, so the total area is $c^2 + 2ab$. One way of computing the area of the large square yields $a^2 + b^2 + 2ab$. A second way yields $c^2 + 2ab$. Since they must be the same,

$$a^2 + b^2 + 2ab = c^2 + 2ab$$

Subtracting $2ab$ from both sides (geometrically this amounts to ignoring the shaded portions of Figs. 14.20 and 14.21), we have

$$a^2 + b^2 = c^2$$

14.2 USE OF THE THEOREM

The Pythagorean theorem can be very valuable for indirect measurement. Suppose, as in the perspective drawing of Fig. 14.22 we want to find the

---

[1]Pythagoras (approximately 580–500 B.C.) was a Greek mystic who traveled widely in the East. He is said to have been so awed by the elegance of his discovery that he sacrificed an ox. He founded a school in what is now Italy, devoted largely to the study of numbers and their role in the universe. His students formed a secret brotherhood, the Pythagoreans, which was eventually considered a threat to the state. This led to Pythagoras being sent into exile, where he died.

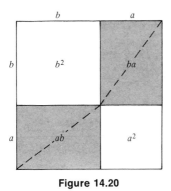

Figure 14.20

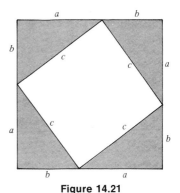

Figure 14.21

distance between towns A and B, but direct measurement between them is impossible because of a mountain. Suppose further that no maps are available, but a point C can be found so that distances CA and CB can be measured and the lines CA and CB form a right angle at C. Specifically, suppose the distances from C to A and C to B turn out to be 21 and 20 miles, respectively. Then the Pythagorean theorem tells us that the square of the distance from C to B is $20^2 + 21^2$, which works out to be $400 + 441 = 841$. Trying a few numbers, we find that $29^2 = 841$, so that the distance from town A to town B is 29 miles. (The square of -29 is also 841, but the distance sought is a positive number.) In this example and all others in this chapter the numbers have been arranged to be integers. Examples in which the numbers do not work out so smoothly are postponed until after the discussion of square roots, which they involve.

As a second example, suppose we must find the distance y from P to Q across a bay when we can easily measure distances over land but not over water. From P we set off in a direction perpendicular to the direction from P to Q until a point R is reached from which line RQ lies entirely over land, as

Figure 14.22

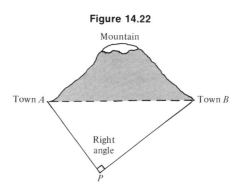

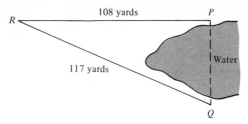

Figure 14.23

shown in Fig. 14.23. Measurement of the land distances shows that *RP* is 108 yards and *RQ* is 117 yards. Since *RP* is perpendicular to *RQ*, triangle *PQR* has a right angle at *P*, and the opposite side, *RQ*, is the longest side of the right triangle. By the Pythagorean theorem we know that the square of 117 is equal to the sum of the squares of *RP* and *PQ*, which are, respectively 108^2 and y^2. Thus $117^2 = 108^2 + y^2$. Since $117^2 = 13,689$ and $108^2 = 11,664$, we have $13,689 = 11,664 + y^2$. Subtracting 11,664 from both sides gives $2,025 = y^2$. A bit of experimenting shows that $45^2 = 2.025$, so the distance *y* is 45 yards.

Exercises 14.2

These exercises have whole number answers.

1   The short sides of a right triangle are 16 and 30. Find the longest side.
2   The short sides of a right triangle are 12 and 35. Find the longest side.
3   The short sides of a right triangle are 24 and 45. Find the longest side.
4   The longest side of a right triangle is 39, and another side is 36. How long is the third side?
5   The longest side of a right triangle is 195, and a second side is 189. How long is the third side?
6   The longest side of a right triangle is 169, and the shortest side is 119. How long is the third side?
7   The three sides of a triangle are 120, 391, and 409. Is this a right triangle? If so, between which two sides is the right angle?
8   One ship is 27 miles due east of a lighthouse, while another ship is 120 miles due north of the same light. How many miles apart are the ships? (Neglect the earth's curvature.)
9   A rectangular lot is 48 yards long and 55 yards wide. How far apart are opposite corners of the lot?
10   A radio transmitting tower is to be supported by cables running from a point on the tower 288 feet above the ground to points on the ground 120 feet from the tower. The ground is level, and the tower is so slender that for this problem it may be considered a pole. How long must each cable be?
11   A 25-foot ladder leans against a wall, touching it at a point 24 feet above the ground, which is level. How many feet must the base of the ladder be moved to lower the point at which the top of the ladder touches the wall by 4 feet?
12   (*a*) Trace the shapes in Fig. 14.24 on a piece of paper and cut them out.
      (*b*) Put pieces *A*, *B*, *C*, and *D* together to form a square.
      (*c*) Make a larger square from pieces *A*, *B*, *C*, *D*, and *E*.
      (*d*) Can you see how this illustrates a case of the Pythagorean theorem?

*14.3 THE STRANGE TALE OF FERMAT'S LAST THEOREM

Examples in this chapter have involved many triples of positive integers a, b, and c for which $a^2 + b^2 = c^2$. To a mathematician, always seeking to generalize his discoveries, it is natural to ask about positive integers a, b, and c for which $a^3 + b^3 = c^3$, $a^4 + b^4 = c^4$, and in general, for which $a^n + b^n = c^n$, where n is an integer greater than 2. This simple question was raised by Fermat, who thought he had settled it, but today it is still not answered, and its history is one of the most interesting stories in mathematics.

Pierre de Fermat (1601–1665) was a member of parliament in Toulouse, France, who paid little attention to mathematics until he was well over thirty. He then became a very great mathematician indeed. His most remarkable contributions were in number theory, which aroused his interest when he read a translation of "Arithmetica," an ancient classic by Diophantus of Alexandria. Fermat had the habit of writing notes in the margin, and with one such note, found after his death, begins the story of "Fermat's last theorem."

On a page of "Arithmetica" devoted to expressing one square as a sum of two others, Fermat wrote, in Latin, "On the other hand it is impossible to separate a cube into two cubes or a fourth power into two fourth powers or

Figure 14.24

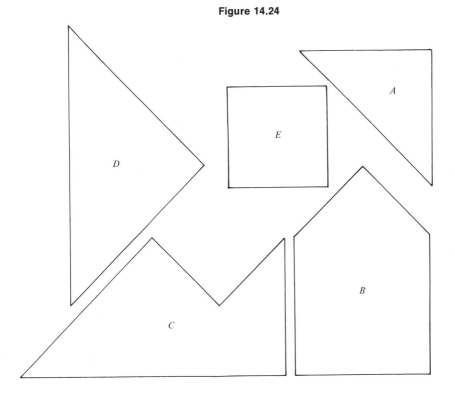

generally any power but a square into powers with the same exponent. I have discovered a truly marvellous proof of this which, however, the margin is too narrow to contain."

Did Fermat really have a proof? Had he made a mistake? To this day this is an open question. Possibly he erred, but two facts must be weighed against this. First, it is known that if positive integers a, b, c, and n are ever to be found for which $a^n + b^n = c^n$, then $n > 25,000$ and a, b, and c each have at least 320,000 digits. Second is Fermat's record of accuracy and honesty. Some of his guesses have been proved wrong (see Prob. 7 of Exercises 6.5), but nothing he ever claimed to have proved has later been disproved.

Fermat's assertion is known as his *last theorem*, because it is the last of his statements to be settled. Strictly, it should not be called a theorem at all until its truth is conclusively established, and since Fermat's work is lost, the issue has remained in doubt. This is not for lack of effort, however; both in 1823 and 1850 the Academy of Science of Paris offered a prize for a correct proof of Fermat's last theorem, as did the Academy of Brussels in 1883. F. F. Wolfskehl of Darmstadt, Germany left 100,000 marks for the same purpose, but its value was reduced to less than a penny by Germany's post–World War I inflation.

In addition to thousands of quack attempts (over 2,000 incorrect papers were published on this topic between 1908 and 1918 alone) a great deal of good work has been stimulated by the problem, but a complete solution seems as far off today as ever.

CHAPTER 15
SQUARE ROOTS

The last step in using the Pythagorean theorem is to find the length of the missing side when the square of that number is known. Sometimes, as in Chap. 14, this can be done by guessing, but in other cases it is not so easy. Here we consider the problem in more detail, because it has interesting connections with our previous work, especially from Chaps. 10 and 12. No new computational or conceptual skills are needed in this chapter, but you will find opportunities to use and solidify earlier ideas in a new context.

15.1 THE PROBLEM AND SOME NOTATION

If we are given a number x, can we find a number y whose square is x? Such a number y is called a *square root* of x. For example, 3 is a square root of 9, and 7 is a square root of 49. Can you find a square root of 25? Of 144? We say "a" square root instead of "the" square root because a number may have more than one square root. For example, 3 is not the only square root of 9, since -3 also has the property that its square is 9. In general, if y is a square root of x, then so is $-y$, since $y^2 = (-y)^2$. Any positive number has two square roots, each the opposite of the other. On the other hand, whether a number is positive or negative, its square is positive, so that a negative number has no square root at all. To overcome this injustice, a whole new family of numbers, called *complex numbers*, has been invented. We shall not go into them here, however, for we shall not deal with square roots of negative numbers.

Since the two square roots of a positive number are opposites of each other, the problem of finding them both is essentially solved if one of them is found. Therefore we concentrate on how to find the *positive* square root, denoted by the sign $\sqrt{}$. For example, $\sqrt{64} = 8$, $\sqrt{25} = 5$, and $\sqrt{\frac{1}{4}} = \frac{1}{2}$. We do not apply this sign to negative numbers, but we do apply it to 0, and $\sqrt{0} = 0$.

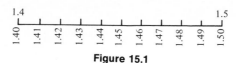

Figure 15.1

The square-root sign acts on whatever number is inside it as a single number. For example, $\sqrt{64 + 36} = \sqrt{100} = 10$; this is *not* the same as $\sqrt{64} + \sqrt{36} = 8 + 6 = 14$.

Exercises 15.1

1 Compute:

(*a*) $\sqrt{169 - 25}$ (*b*) $\sqrt{169} - \sqrt{25}$ (*c*) $\sqrt{15^2 + 8^2}$

(*d*) $\sqrt{15^2 + 8^2}$ (*e*) $\sqrt{\sqrt{9} + \sqrt{169}}$

2 Compute:

(*a*) $\sqrt{4^2}$ (*b*) $\sqrt{(-4)^2}$ (*c*) $\left(\sqrt{4}\right)^2$

(*d*) $\left(-\sqrt{4}\right)^2$ (*e*) $-\left(\sqrt{4}\right)^2$

15.2 SUCCESSIVE APPROXIMATIONS

Of the many ways to compute square roots, the simplest in principle is the ancient method of successive approximations, which is based on repeated guessing. You probably used this method, without giving it a fancy name, when you did the problems in Chap. 14. To illustrate the method, we shall use it on $\sqrt{2}$.

Clearly $\sqrt{2}$ lies between 1 and 2, since $1^2 = 1$, and $2^2 = 4$. If we guess $1\frac{1}{2}$, we find $\left(1\frac{1}{2}\right)^2 = \left(\frac{3}{2}\right)^2 = \frac{9}{4}$, which is more than 2, so $1\frac{1}{2} > \sqrt{2}$. We now know that $\sqrt{2}$ lies between 1 and $1\frac{1}{2}$. If we guess $1\frac{1}{4}$, we find its square is $\left(\frac{5}{4}\right)^2 = \frac{25}{16} < 2$, and so $1\frac{1}{4} < \sqrt{2}$. We now know $\sqrt{2}$ is between $1\frac{1}{4}$ and $1\frac{1}{2}$. From here on it is easier to use decimals. We might next guess 1.4, but since $1.4^2 = 1.96 < 2$, 1.4 is too small. With each step we have narrowed the search. We still do not know $\sqrt{2}$ precisely, but at this stage we can be sure that $\sqrt{2}$ is between 1.4 and 1.5. To carry the process further, we subdivide the interval from 1.4 to 1.5 as in Fig. 15.1 and continue guessing.

Eventually we find $1.41^2 = 1.9881$ and $1.42^2 = 2.0164$, which shows that $\sqrt{2}$ is between 1.41 and 1.42. Now we could subdivide that interval as in Fig. 15.2 and continue the process. If we did so we could eventually conclude that $\sqrt{2}$ lies between 1.414 and 1.415, since $1.414^2 = 1.999396$, and $1.414^2 = 2.002225$.

When will this process end? Unfortunately, never.[1] No matter how far the

---

[1] In 1971 a mathematician at Columbia University used a high-speed computer to calculate more than a million decimal places of $\sqrt{2}$.

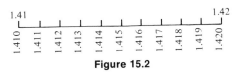

Figure 15.2

process is carried, it never reaches $\sqrt{2}$ with perfect accuracy, but it certainly gives us approximations which are close enough for any practical problem.

Newton invented an improved version of this method.[2] To find the square root of a given number n, begin by taking a guess, and divide n by your guess. If you are lucky enough to guess $\sqrt{n}$ perfectly, the quotient will be the same as the number you guessed. Otherwise the square root you seek will lie between your guess and the quotient. Therefore, average your guess and the quotient to get an improved guess. Repeat this process until the desired accuracy is reached.

To illustrate, suppose we want to find $\sqrt{3}$ to the nearest hundredth. Clearly $\sqrt{3}$ lies somewhere between 1 and 2, and so for a first, rough guess we try 1.5. Now $3 \div 1.5 = 2$, so $1.5 < \sqrt{3} < 2$. For our next guess we use the average of 1.5 and 2, namely 1.75.

$$
\begin{array}{r}
1.71\cdots \\
1.75\overline{)3.000000} \\
1\,75 \\
\overline{1\,250} \\
1\,225 \\
\overline{250}
\end{array}
$$

(We do not complete the division but carry it only to a decimal place where divisor and quotient differ. This lets us move quickly to a better guess and has little effect on its accuracy.) Averaging 1.75 and 1.71, we next try 1.73.

$$
\begin{array}{r}
1.734\cdots \\
1.73\overline{)3.000000} \\
1\,73 \\
\overline{1\,270} \\
1\,211 \\
\overline{590} \\
519 \\
\overline{710} \\
692
\end{array}
$$

[2]Sir Isaac Newton (1642–1727) was the son of an English farmer. Because of his great scientific talents, he went to Cambridge. There he became interested in mathematics and became one of the inventors of that immense and subtle part of the subject, the calculus. He showed that a gravitational force would cause planets to move in elliptical orbits about the sun and worked extensively in other areas of physics, especially optics, as well as in mathematics.

From this we see that, to the nearest hundredth, $\sqrt{3}$ is 1.73. For still more accuracy we could continue the process, averaging 1.73 and 1.734 to get our next guess of 1.732.

Successive approximations are not the fastest ways to calculate square roots, but the methods are simple in principle and keep attention firmly on the main question. (They can also be a good source of computation practice.)

Exercises 15.2

1 (a) Use successive approximations to find $\sqrt{8}$ to the nearest thousandth.
   (b) Compare your answer from part (a) with the approximate value of $\sqrt{2}$ found in the text. Can you explain this?
2 (a) By successive approximations, find $\sqrt{5}$ to the nearest thousandth.
   (b) Multiply your answer from part (a) by 1.414, which is $\sqrt{2}$ to the nearest thousandth, and round off the product to the nearest thousandth. What square root does $\sqrt{5} \cdot \sqrt{2}$ equal?
3 Find a way to use the multipliers in Chap. 12 to find square roots approximately. (This is so delightfully simple that you will be surprised. We shall consider the question in more detail later.)
4 Find the longest side of a right triangle to the nearest $\frac{1}{10}$ foot if the two shorter sides are:
   (a) 6 and 7 feet long   (b) 8 and 11 feet long
   (c) 3.4 and 7.5 feet long   (d) 57 and 72 feet long
5 A baseball diamond is a square 90 feet on an edge. If a catcher standing on home plate wants to throw a man out at second, how far must he throw if the second baseman is standing on the bag? (Approximate your answer to the nearest $\frac{1}{10}$ foot.)
6 The longest side of a right triangle is 100 feet long. How long is the third side if a second side is:
   (a) 11 feet long   (b) 27 feet long   (c) 50 feet long
7 If you walk 3 miles north and then 5 miles east, how far are you from your starting point? (Assume you are not anywhere near the North or South Pole.)
8 A rectangular television screen measures 19 inches between diagonally opposite corners. If the screen is 15 inches long, how high is it?

*15.3 IRRATIONAL NUMBERS

Is there a better way to express $\sqrt{2}$ than as an infinite decimal whose digits must be laboriously determined one at a time? Fraction form, for example, would be much handier. However, the ancient Greeks discovered to their dismay that there is no such fraction.

At first you may find this absurd. The main reason for expanding the concept of number to include fractions was to allow us to measure quantities, such as lengths, which are not a whole number of units, but now you are told that the length of the long side of the triangle in Fig. 15.3 cannot be represented by a fraction. The distinction is theoretical rather than practical, since all measurement is approximate, and there are fractions which approximate $\sqrt{2}$ as closely as we please. But we shall now show that there is no fraction whose square is precisely 2. Following the Greeks, we use indirect reasoning. You may want to reread the discussion of indirect reasoning in Sec. 4.4 before going on.

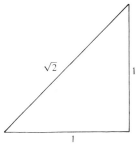

Figure 15.3

Clearly there are just two possibilities: either there is a fraction whose square is 2 or there is not. We shall assume there is such a fraction and then show that this leads us to contradict something we are sure is true. If there is a fraction whose square is 2, it can surely be expressed in lowest terms, for any fraction can be so expressed. The gist of our reasoning will be to show that no fraction which equals $\sqrt{2}$ could possibly be in lowest terms, and that amounts to showing that there is no fraction which equals $\sqrt{2}$.

Suppose that, in lowest terms, p/q is a fraction whose square is 2. Here p and q are whole numbers, and since the fraction is in lowest terms, p and q have no prime factors in common. Now $(p/q)^2 = 2$ means that $p^2 = 2q^2$. Since q is an integer, so is q^2, and therefore $2q^2$ is an even integer. Since $2q^2 = p^2$, p^2 is an even integer. Is p even or odd? If p were odd, p^2 would be odd, too, so p is even. That means p is twice some integer r, so $p^2 = (2r)^2 = 4r^2$. Thus $4r^2 = 2q^2$; and, dividing by 2, $2r^2 = q^2$. By the same reasoning, we find q is even ($2r^2$ is even, so q^2 is even, which means q is even). But if p and q were both even, then p/q was not in lowest terms, as a factor of 2 could have been canceled. The only way out of this is to drop the assumption which led us here, that $\sqrt{2}$ can be expressed as a fraction.

True, we have deliberately contrived a contradiction. Had we not done so, it would have lurked, temporarily undetected; it would not have gone away.

A number which can be expressed as a fraction is said to be *rational*, a word related to ratio. (The *ratio* of two numbers is just their quotient, though it is sometimes written with a colon instead of a fraction line. For example, the ratio 3 to 2 is written $3 : 2$ and means $\frac{3}{2}$.) Virtually all numbers used in daily life are rational, but $\sqrt{2}$ is not. Numbers which are not rational are said to be *irrational*.

There are many irrational numbers, including $\sqrt{3}$, $\sqrt{5}$, and in general all square roots which do not come out even. Many of the most interesting and important numbers in mathematics, such as π, are irrational.

Decimals bring out another aspect of irrational numbers. The decimal form of a rational number has one of the following to the right of the decimal point:

1. Only zeros

2. A finite number of nonzero digits

3. An infinite number of nonzero digits which eventually fall into a cyclically repeating pattern (as seen in Chap. 10)

Now consider

$$0.20200200020000200000 2 \cdots$$

where the dots indicate that the pattern of 2s separated by increasing strings of 0s continues forever. This pattern is not cyclic, and so the infinite decimal does not fall into any of the three classes above. Therefore it does not represent a rational number. Any infinite decimal whose digits never settle into a periodic pattern represents an irrational number. They are easy to make up, as these examples show:

$$0.123456789101112131415161718192021 22 \cdots$$

$$0.8182838485868788898108118 \cdots$$

$$0.32332233322233332222 \cdots$$

*15.4 CONTINUED FRACTIONS

Continued fractions are off the beaten track, but they are ideal for classroom use, because they combine an interesting and open-ended exploration with an excellent review of some basic techniques for handling fractions.

We begin with the series:

$$1 + \frac{1}{2}, \quad 1 + \cfrac{1}{2 + \cfrac{1}{2}}, \quad 1 + \cfrac{1}{2 + \cfrac{1}{2 + \cfrac{1}{2}}}, \quad 1 + \cfrac{1}{2 + \cfrac{1}{2 + \cfrac{1}{2 + \cfrac{1}{2}}}}, \cdots$$

These may look like the work of a fevered brain. What shall we do with them? A good start would be to evaluate them. The first is clearly $\frac{3}{2}$. How about the second? Starting at the bottom, we write $2 + \frac{1}{2}$ as $\frac{5}{2}$, to make it

$$1 + \cfrac{1}{\frac{5}{2}}$$

Since dividing by $\frac{5}{2}$ is the same as multiplying by $\frac{2}{5}$,

$$\frac{1}{\frac{5}{2}} \text{ is just } 1 \cdot \frac{2}{5}, \quad \text{and} \quad 1 + \cfrac{1}{\frac{5}{2}} \text{ is } 1 + \frac{2}{5} \quad \text{or} \quad \frac{7}{5}$$

More complicated examples are done by the same method, working from the bottom up and repeatedly adding and dividing. Let's do the next one together; fill in the blanks at each step.

$$1 + \cfrac{1}{2 + \cfrac{1}{2 + \cfrac{1}{2}}} = 1 + \cfrac{1}{2 + \cfrac{1}{\cfrac{1}{2} + \cfrac{1}{2}}} = 1 + \cfrac{1}{2 + \cfrac{1}{2}}$$

$$= 1 + \cfrac{1}{2 + \cfrac{1}{5}} = 1 + \cfrac{1}{\cfrac{2}{5} + \cfrac{2}{5}} = 1 + \cfrac{1}{\cfrac{1}{5}} = 1 + \cfrac{1}{12}$$

$$= \cfrac{}{12} + \cfrac{5}{12} = \cfrac{}{12}$$

Try the next one yourself. If you do it correctly, you should get $\frac{41}{29}$.

The answers so far, $\frac{3}{2}, \frac{7}{5}, \frac{17}{12}$, and $\frac{41}{29}$, fit a pattern. Can you find it? (*Hint*: If you know one, how could you find the denominator of the next?) Now can you find the rule that determines its numerator? According to the pattern, what should be the value of

$$1 + \cfrac{1}{2 + \cfrac{1}{2 + \cfrac{1}{2 + \cfrac{1}{2 + \cfrac{1}{2}}}}}$$

Check your prediction by actually working this out. A remarkable aspect of these continued fractions emerges if we square their values. Complete

$$\left(\tfrac{3}{2}\right)^2 = \qquad\qquad \left(\tfrac{7}{5}\right)^2 =$$

$$\left(\tfrac{17}{12}\right)^2 = \qquad\qquad \left(\tfrac{99}{70}\right)^2 =$$

$$\left(\tfrac{41}{29}\right)^2 =$$

What do you notice about these squares? They are alternately above and below a certain fixed number, which they seem to be approaching. What is that number? That means the numbers $\frac{3}{2}, \frac{7}{5}, \frac{17}{12}, \frac{41}{29}, \frac{99}{70}, \cdots$ are alternately above and below $\sqrt{2}$ but getting closer to it. Carrying the series further, we can find fractions which are as close as we please to $\sqrt{2}$, though none is exactly $\sqrt{2}$. This is an interesting idea, since the series of fractions is so easy to extend. You have probably already found the following rule, perhaps in another form. To extend the list $\frac{3}{2}, \frac{7}{5}, \frac{17}{12}, \frac{41}{29}, \frac{99}{70}, \cdots$ to a new fraction, add the numerator and denominator of the last one to get the new denominator, and add the numerator to twice the denominator to get the new numerator. Thus, in this case the next fraction is

$$\frac{2 \cdot 70 + 99}{99 + 70} = \frac{239}{169}$$

These discoveries beg for more questions to be asked. Think of some that would be natural at this stage, and how you might try to answer them. The exercises below give a few.

Exercises 15.4

1 Do the continued fractions

$$1 + \frac{1}{3}, \qquad 1 + \cfrac{1}{3 + \frac{1}{3}}, \qquad 1 + \cfrac{1}{3 + \cfrac{1}{3 + \frac{1}{3}}}, \cdots$$

approach $\sqrt{3}$ the way those discussed above tend to $\sqrt{2}$? Justify your answer,

2 (*a*) Evaluate

$$2 + \frac{1}{4}, \qquad 2 + \cfrac{1}{4 + \frac{1}{4}}, \qquad 2 + \cfrac{1}{4 + \cfrac{1}{4 + \frac{1}{4}}}, \qquad 2 + \cfrac{1}{4 + \cfrac{1}{4 + \cfrac{1}{4 + \frac{1}{4}}}}$$

(*b*) Find a rule like that discussed in the text which allows you to predict the values of the continued fractions

$$2 + \cfrac{1}{4 + \cfrac{1}{4 + \cfrac{1}{4 + \frac{1}{4}}}} \qquad \text{and} \qquad 2 + \cfrac{1}{4 + \cfrac{1}{4 + \cfrac{1}{4 + \cfrac{1}{4 + \frac{1}{4}}}}}$$

(*c*) Square your answers to part (*a*), and find out what square root the continued fractions approximate.

3 Repeat the instructions from Prob. 2 for:

(*a*) $3 + \frac{1}{6}, \qquad 3 + \cfrac{1}{6 + \frac{1}{6}}, \qquad 3 + \cfrac{1}{6 + \cfrac{1}{6 + \frac{1}{6}}}, \qquad 3 + \cfrac{1}{6 + \cfrac{1}{6 + \cfrac{1}{6 + \frac{1}{6}}}}, \cdots$

(*b*) $4 + \frac{1}{8}, \qquad 4 + \cfrac{1}{8 + \frac{1}{8}}, \qquad 4 + \cfrac{1}{8 + \cfrac{1}{8 + \frac{1}{8}}}, \qquad 4 + \cfrac{1}{8 + \cfrac{1}{8 + \cfrac{1}{8 + \frac{1}{8}}}}, \cdots$

4 On the basis of the above work, what square roots do you think can be approximated by these continued fractions?

(a) $10 + \dfrac{1}{20}$, $\quad 10 + \dfrac{1}{20 + \dfrac{1}{20}}$, $\quad 10 + \dfrac{1}{20 + \dfrac{1}{20 + \dfrac{1}{20}}}$, $\ldots$

(b) $14 + \dfrac{1}{28}$, $\quad 14 + \dfrac{1}{.28 + \dfrac{1}{28}}$, $\quad 14 + \dfrac{1}{28 + \dfrac{1}{28 + \dfrac{1}{28}}}$, $\ldots$

(c) generalize

*15.5 INTRODUCTION TO RATIONAL EXPONENTS

We first used exponents to express repeated multiplication; later we generalized to negative exponents. Now we generalize in another direction—to fractional exponents. As before, multipliers point the way.

Multipliers are adders for exponents, and the adder in Sec. 2.4 works for fractions as well as integers. We saw in Chap. 12 that a typical multiplier can be expressed with exponents as shown in Fig. 15.4:

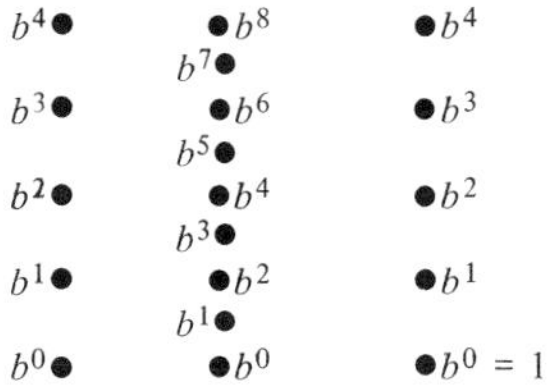

$$
\begin{array}{ccc}
b^4 \bullet & \begin{array}{l} \bullet b^8 \\ b^7 \bullet \end{array} & \bullet b^4 \\[4pt]
b^3 \bullet & \begin{array}{l} \bullet b^6 \\ b^5 \bullet \end{array} & \bullet b^3 \\[4pt]
b^2 \bullet & \begin{array}{l} \bullet b^4 \\ b^3 \bullet \end{array} & \bullet b^2 \\[4pt]
b^1 \bullet & \begin{array}{l} \bullet b^2 \\ b^1 \bullet \end{array} & \bullet b^1 \\[4pt]
b^0 \bullet & \bullet b^0 & \bullet b^0 = 1
\end{array}
$$

Figure 15.4

Where should $b^{1/2}$ fit here? Evidently it belongs midway between b^0 and b^1. But what number is $b^{1/2}$? Since we have a multiplier, $b^{1/2}$, whatever it is, has the property that $b^{1/2}b^{1/2} = b^1$. With that in mind, we define $b^{1/2}$ to be $\sqrt{b}$ for *any* positive number b. This agrees with earlier observations about the multipler. Specifically, to square a number locate it in an outer column; its square is at the same level in the center column. Therefore, a number in an outer column is the positive square root of the number at the same height in the center column.

Note that our definition of $b^{1/2}$ has the property that $b^{1/2}b^{1/2} = b^1$, which extends the rule, found for *integers* m and n,

$$b^m b^n = b^{m+n}$$

to the case where $m = n = \frac{1}{2}$. Further, since $b^{1/2}b^{1/2} = (b^{1/2})^2$, it also extends the rule

$$(b^m)^n = b^{mn}$$

which was also found for integers m and n, to the case where $m = \frac{1}{2}$ and $n = 2$.

Exercises 15.5

1 Evaluate:
   (*a*) $4^{1/2}$ (*b*) $9^{1/2}$ (*c*) $1^{1/2}$ (*d*) $49^{1/2}$
2 Compare $(9 + 16)^{1/2}$ and $9^{1/2} + 16^{1/2}$. What does this example show?
3 Evaluate:

   (*a*) $\left(\frac{1}{4}\right)^{1/2}$ (*b*) $\left(\frac{1}{9}\right)^{1/2}$ (*c*) $0.01^{1/2}$

   (*d*) $\frac{1}{4^{1/2}}$ (*e*) $\frac{1}{9^{1/2}}$ (*f*) $\frac{1}{100^{1/2}}$

4 Evaluate:

   (*a*) $\left(\frac{4}{9}\right)^{1/2}$ (*b*) $\frac{4^{1/2}}{9^{1/2}}$ (*c*) $\left(\frac{36}{49}\right)^{1/2}$ (*d*) $\frac{36^{1/2}}{49^{1/2}}$
   (*e*) generalize
5 Evaluate:
   (*a*) $(100^{1/2} + 36^{1/2})^{1/2}$ (*b*) $(289^{1/2} - 64^{1/2})^{1/2}$
6 Use the multiplier from Fig. 12.12 (p. 121) to estimate:
   (*a*) $14.42^{1/2}$ (*b*) $25.54^{1/2}$ (*c*) $80.15^{1/2}$

   In the following exercises you will generalize the above work to other rational exponents.

7 (*a*) $4^3 =$ (*b*) $(4^3)^{1/2} =$ (*c*) $(4^{1/2})^3 =$ (*d*) $4^1 \cdot 4^{1/2} =$
   (*e*) On the basis of parts (*b*), (*c*), and (*d*), give a definition of $4^{3/2}$ that conforms to the rules
   $4^m 4^n = 4^{m+n}$ and $(4^m)^n = 4^{mn}$ when either m or n is $\frac{1}{2}$.
   (*f*) Does your definition of $4^{3/2}$ also satisfy the rule $4^m \div 4^n = 4^{m-n}$ when $m = 2$ and $n = \frac{1}{2}$?
   Check it!
8 Following the thinking from Prob. 7, evaluate:
   (*a*) $9^{3/2}$ (*b*) $100^{3/2}$ (*c*) $289^{3/2}$

   (*d*) $\left(\frac{1}{4}\right)^{3/2}$ (*e*) $\left(\frac{16}{25}\right)^{3/2}$ (*f*) $(0.09)^{3/2}$

9 As a check on the concept of fractional powers that is emerging, evaluate:

   (*a*) $(4^4)^{1/2}$ (*b*) $(4^{1/2})^4$
   (*c*) If our ideas fit earlier patterns, then the answers for parts (*a*) and (*b*) should both yield
   $4^{4/2}$. Since $\frac{4}{2} = 2$, this is easy to check. Are the answers what they should be?
10 Extend your patterns still further by figuring out how much these are.
    (*a*) $4^{5/2}$ (*b*) $9^{5/2}$ (*c*) $100^{5/2}$ (*d*) $4^{7/2}$

    (*e*) $\left(\frac{1}{4}\right)^{5/2}$ (*f*) $\left(\frac{1}{9}\right)^{5/2}$ (*g*) $\left(\frac{16}{9}\right)^{5/2}$

11 We now extend the above considerations in another direction. How shall we define $8^{1/3}$? Note
    that if the rule $(a^m)^n = a^{mn}$ holds, then we should have $(8^{1/3})^3 = 8^{3/3} = 8^1 = 8$, so evidently $8^{1/3}$ is a
    number whose cube is 8. This number is called a *cube root* of 8. Can you find it by trial? Cube
    roots are generally more tedious to calculate than square roots, but we do not go into that.
    What are:
    (*a*) $27^{1/3}$ (*b*) $125^{1/3}$ (*c*) $1{,}000^{1/3}$
12 If we reason as above, $8^{2/3}$ should be $(8^{1/3})^2$. How much are $27^{2/3}$ and $125^{2/3}$?

13 If our approach so far makes sense, then the second power (square) of any number should be the same as the sixth power of its cube root; that is, we should have

$$(8^{1/3})^6 = 8^{6/3} = 8^2 \quad \text{and} \quad (27^{1/3})^6 = 27^{6/3} = 27^2$$

and so forth. Check these and a few like them to see if they work the way they should.

Our ideas can be extended to negative rational exponents by simply combining ideas previously developed for negative integer exponents and positive fractional exponents. The following problems will bring that out.

14 Evaluate:
(a) 4^{-1} (b) $(4^{-1})^{1/2}$ (c) $(4^{1/2})^{-1}$
(d) Use your answers to parts (b) and (c) to define $4^{-1/2}$.
(e) To check that your definition is reasonable, see whether, with it,

$$4^{-\frac{1}{2}} \cdot 4^{-\frac{1}{2}} = 4^{\left(-\frac{1}{2}\right) + \left(-\frac{1}{2}\right)} = 4^{-1}$$

(f) As a further check, compute $4^1 \cdot 4^{-\frac{1}{2}}$. Is this the same as $4^{1 + \left(-\frac{1}{2}\right)} = 4^{\frac{1}{2}}$?
(g) As a final check, compute $(4^{-\frac{1}{2}})^{-1}$. It should be the same as $4^{\left(-\frac{1}{2}\right)(-1)}$. Is it?

15 Use reasoning like that in Prob. 13 to evaluate:

(a) $9^{-1/2}$ (b) $25^{-1/2}$ (c) $\left(\dfrac{1}{2}\right)^{-1/2}$ (d) $\left(\dfrac{49}{64}\right)^{-1/2}$

(e) $16^{-3/2}$ (f) $\left(\dfrac{9}{4}\right)^{-3/2}$ (g) $\left(\dfrac{27}{64}\right)^{-2/3}$

CHAPTER 16
ANGLES

The idea of angle underlies any study of shape and form. In this chapter we consider what angles are, how they are measured, and some basic relationships among them. Later, when we discuss polygons and polyhedra, we shall use these ideas extensively.

16.1 WHAT IS AN ANGLE?

There are many ways to define the concept of angle, most of them too formal for youngsters. Here is an intuitive approach.

If you ask someone what an angle is, he will probably draw something like Fig. 16.1. By itself this does not define the idea, but it is a start. Figure 16.2 shows an enlargement of Fig. 16.1; is it a larger angle? The person who drew Fig. 16.1 would doubtless reply that the angles in the two figures are the same size. Then what must be done to Fig. 16.1 to make it a larger angle? Just swing one of the line segments away from the other, as in Fig. 16.3.

Evidently the concept of angle is pictured by the relative *directions* of the line segments rather than their lengths. With this in mind we define an angle to be a rotation. Of course to use this, one must understand what a rotation is, but most people grasp this idea intuitively.

Figure 16.1   Figure 16.2

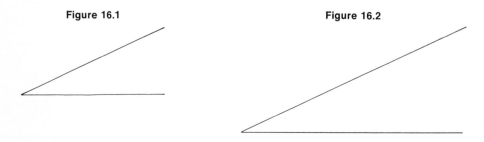

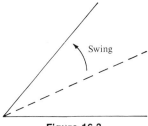

Figure 16.3

Whenever anything rotates or turns, angles are involved even if no pictures are drawn. Often, however, we want to picture an angle, and for that we use a version of Fig. 16.1 modified by a curved arrow from one line segment to the other. We think of the rotation as moving one line segment to the other as indicated by the arrow. Figure 16.4 shows three different angles, all of which are rotations sending line segments S_1 to S_2.

This way of picturing angles, though widely used, is not entirely standard; sometimes pictures like Fig. 16.1, with no curved arrows, are used to represent angles. In such cases it is safe to assume that the direction of rotation is irrelevant and that the angle involves the least possible rotation sending one line segment to the other.

16.2 HOW ARE ANGLES MEASURED?

To measure anything we must choose a unit. A natural unit of angular measure is the complete turn, pictured in Fig. 16.5, and all other angles can be expressed in terms of this unit. A half turn (Fig. 16.6) is called a *straight angle*, and a quarter turn (Fig. 16.7) is called a *right angle*. The angular unit used most often in practical applications is the *degree* (the small circle denotes degrees) which

Figure 16.4

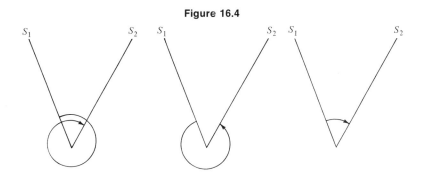

Figure 16.5

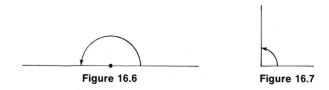

Figure 16.6   **Figure 16.7**

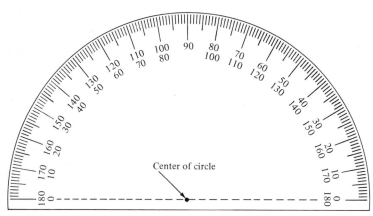

Figure 16.8

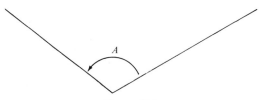

Figure 16.9

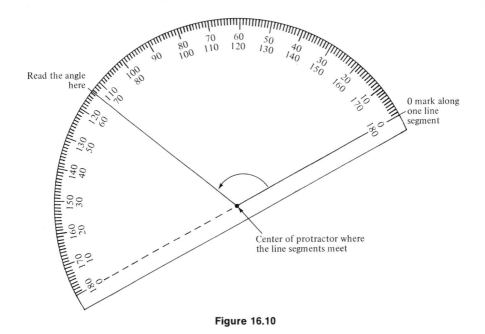

Read the angle here

0 mark along one line segment

Center of protractor where the line segments meet

Figure 16.10

is $\frac{1}{360}$ of a complete turn. A complete turn therefore measures 360°. An angle of 720° is two complete turns, whereas a straight angle is 180°. Sometimes angles in one direction (say, counterclockwise) are considered positive, while those in the other direction are negative, but we shall not follow that convention.

Lately it has been fashionable to distinguish elaborately between an angle and its measure; one speaks not of a "20° angle," but of an "angle whose measure is 20°." This distinction, like that between number and numeral, is logically correct but often labored unnecessarily. We shall speak freely of a "20° angle," confident this will cause no confusion.

The basic tool for measuring angles is a *protractor*; it does for angles what a ruler does for lengths. Essentially a protractor is just a circle marked off in 360 equal parts, though in practice it is common to use only half the circle, as in Fig. 16.8. A good protractor is made of clear plastic, is large enough to read easily, and has its zero marks away from the straight edge, so that if this edge is uneven, from manufacture or misuse, accuracy is not affected. Most protractors, like that in Fig. 16.8, are labeled with two scales, one for clockwise angles and the other for counterclockwise angles.

To measure an angle represented by a picture, as in Fig. 16.9, place the protractor with its center where the two line segments meet and its 0 mark so it lies along the segment, as in Fig. 16.10. Then read the number where the other segment crosses the edge of the protractor. In Fig. 16.10 angle A is measured at about $113\frac{1}{2}°$. If the angle is larger than 180°, how can a protractor be used to measure it, since the protractor's scales only go up to 180°? This will be covered in Prob. 6 of the exercises.

   Just as a ruler can be used to "construct" points a given distance apart, a protractor can be used to make line segments whose directions differ by a given angle. This is usually called constructing the angle. For example, to construct a 36° angle, first mark the point where the line segments will meet, and place the protractor with its center at that point. Next mark two points at the edge at scale numbers which differ by 36; 0 and 36 are as good as any, but other pairs, such as 10 and 46, will also do. Draw line segments from the center to each of the marks on the edge and complete the job by putting in a curved arrow to show the angle's size and direction.

Exercises 16.2

1 How many complete turns are there in an angle of:
   (*a*) 1,440°   (*b*) 5,400°   (*c*) 0°   (*d*) 810°
   (*e*) 900°   (*f*) 45°   (*g*) 60°   (*h*) 120°
   (*i*) 300°
2 How many degrees are there in:
   (*a*) One-tenth of a complete turn?
   (*b*) One-twelfth of a complete turn?
   (*c*) Three-fifths of a complete turn?
   (*d*) Two-thirds of a complete turn?
   (*e*) No turn at all?
   (*f*) One-seventh of a complete turn?
   (*g*) Eleven complete turns?
   (*h*) $9\frac{3}{4}$ complete turns?
3 (*a*) Through how many degrees does the small (hour) hand of a clock turn in an hour?
   (*b*) Through how many degrees does the big (minute) hand of a clock turn in an hour?
   (*c*) Through how many degrees does the second hand of a clock turn in an hour?
4 At noon the big and little hands of the clock coincide. Through how many degrees must the big hand turn before it next meets the little hand?
5 To one who stands still on the earth, the sun appears to circle the earth at a steady rate. The time of one complete rotation is called a day. How many degrees does the sun appear to move through the sky in 1 hour?
6 Measure these angles with a protractor:

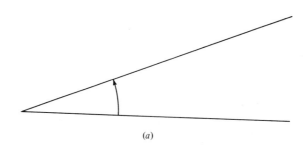

(*a*)

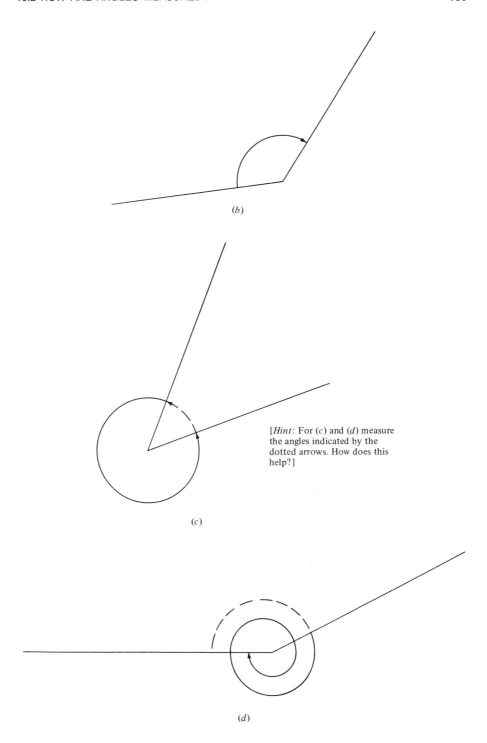

(b)

[*Hint*: For (c) and (d) measure the angles indicated by the dotted arrows. How does this help?]

(c)

(d)

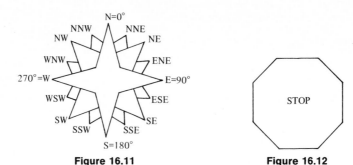

Figure 16.11   **Figure 16.12**

7 Use a protractor to construct angles of:
   (a) 45°   (b) 60°   (c) 37°   (d) 11°   (e) 90°
   (f) 135°   (g) 124°   (h) 270°   (i) 283°   (j) 427°
   (k) 5,047°

8 An ignorant ruler is said to have refused to allow a shipment of English machinery to enter his country because it was marked "1,900 revolutions per minute." Perhaps the shipment would have been accepted if instead it had been labeled in degrees per minute. How many degrees per minute should it have been labeled? Would students in that country be pleased?

In navigation one who is heading due north is said to be on a course of 0°, and all other courses are measured in degrees from north. Thus one who is heading due east is on a course of 90°, and headings of south and west are 180° and 270°, respectively. The compass rose in Fig. 16.11 may help with the following questions.

9 What points on the compass rose correspond to courses of:
   (a) 45°   (b) 315°   (c) 247½°   (d) 112½°

10 What headings in degrees correspond to:
   (a) NW   (b) SW   (c) NNW   (d) ENE
   (e) Halfway between SW and WSW

11 If you are on a course of 20°, what will be your heading in degrees after:
   (a) A 15° turn to the right   (b) A 15° turn to the left
   (c) A 90° turn to the right   (d) A 90° turn to the left
   (e) A 180° turn   (f) A 360° turn
   (g) A 450° turn to the left

12 A course of 60° takes one from Banning Pass to Lake Havasu City. What course takes one from Lake Havasu City to Banning Pass?

13 You are heading due north, but you turn 17° to the right, then 39° to the left. Now what is your heading?

14 If you begin facing due north and make five successive turns of 75° to the left, what direction will you be facing?

15 Suppose you drive around a rectangular block, making four right-hand turns. Before you make your first turn you were on a heading of 219°. What were your headings as you traveled the other three sides of the block?

16 I make three equal turns to the right and end up facing the same direction as when I began. If each turn is more than 90° but less than 150°, how many degrees in each?

17 An octagonal stop sign (Fig. 16.12) blew over in a storm, and an ant from a nearby colony was sent to scout it. He began at the middle of one side of the sign and walked completely around the edge of the sign.
   (a) By the time he returned to his starting point, through how many degrees had he turned in all?

(*b*) How many degrees did he turn at each corner of the sign if they were all alike?

18 (*a*) Measure each of the angles indicated in the triangle in Fig. 16.13.

(*b*) Add the three answers from part (*a*).

(*c*) Repeat parts (*a*) and (*b*) as carefully as you can with at least three more triangles. What do you notice about the sums of the angles?

16.3 EUCLIDEAN AND NON-EUCLIDEAN GEOMETRIES

You probably found in Prob. 18 that for each triangle you tried the sum of the angles is about 180°. Of course this figure is only approximate, since no measurement is perfectly accurate, but if you missed it by more than 4°, you were either very sloppy or made an error, such as not lining up the protractor correctly or reading the wrong scale. Measuring the angles of triangles is an excellent classroom activity, since the practice is combined with a discovery and the teacher can quickly spot those who need help because they do not always come out with about 180°. The teacher should also look out for students who always seem to get exactly 180°. Consciously or not, they may be "fudging" their results to make them right.

Do the angles of every triangle add up to precisely 180°? This question sounds simple, but the answer is not known. Whenever the measurements have been carefully made, the sum of the angles of any triangle has differed from 180° by less than might be accounted for by errors of measurement (remember, all measurement is approximate), so for all practical purposes we may assume that the angles of any triangle total 180°. The theoretical question is not so easily disposed of, however. To see why, consider a situation that might have occurred a few centuries ago when it was believed that the earth was flat. Suppose at that time you could measure angles quite accurately. You measure a few triangles in the classroom, and in each case the angles total

Figure 16.13

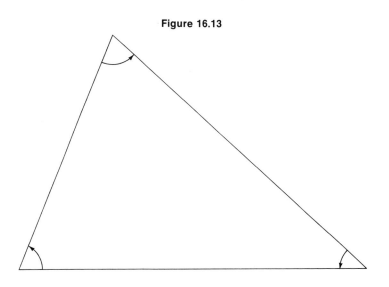

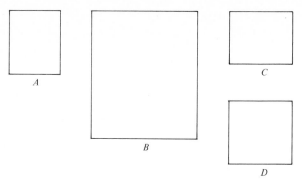

Figure 16.14

180°, at least within the limits of measurement, but now you measure the angles of a much larger triangle formed by ships several miles apart at sea. To your surprise, the angles of this large triangle add up to more than 180°, and the difference cannot be explained away by the errors of measurement. Further, the larger the triangle, the more the sum of the angles exceeds 180°.

This is no mystery to one who realizes that the earth is round. The curvature of the earth makes it impossible to draw truly straight lines on it, and so what we thought was a triangle (a figure with three straight sides) was only a spherical approximation to it.

Could a similar situation exist in three dimensions? What if we tried to measure the angles of a huge interplanetary or even interstellar triangle? Here the same problem reappears in more subtle form. From one corner of this triangle we may sight the other two and measure the angle they appear to form. This is really the angle formed by light beams reaching us from the other two vertices, but we cannot be certain that the path of a light beam fits our concept of a straight line!

For over 2,000 years it was generally believed that the angles of any triangle total precisely 180°, because that conclusion appears in Euclid's "Elements," a work so highly esteemed that its statements were for a long time regarded as orthodoxy.[1] It was believed that any concept of space other than Euclid's must be inherently self-contradictory, so that Euclidean geometry was the only one possible. But in the nineteenth century it was shown that some views of space other than Euclid's are just as free of contradiction, and with this there arose a lively interest in *non-Euclidean geometries*, a topic which is now known to have been discussed in Euclid's time. In some of these geometries the angles of a triangle do not add up to 180°, but the discrepancy depends on the size of the triangle. To this day it is not known whether physical space is better described by Euclidean geometry or by one of the non-Euclidean geometries. We can be sure, however, that for triangles less than a few

---

[1]See note on Euclid and the "Elements" on page 38.

million miles on edge, the discrepancy is unobservable. We are therefore justified in using Euclidean geometry in all practical situations, but we should realize that it is not the only geometry possible.

16.4 SIMILARITY AND CONGRUENCE

Which of the objects in Fig. 16.14 are alike?

It all depends, you might say, what you mean by "alike." They are all alike in that they all have four sides, appear on the same page, and are rectangles. Yet some are more alike than others. D is a square, its length and width being equal. That makes it different from A, B, and C, all of which have the same shape. A and C also have the same size, while B is twice as long and twice as wide as either A or C.

Objects which have the same shape (like A, B, and C) are said to be *similar*. Objects like A and C which have both the same shape and the same size are said to be *congruent*. Mass-produced items provide plentiful examples of objects which are congruent for all practical purposes.

Objects which are similar are scale replicas of each other. For example, if a model boat is built to a scale of $\frac{1}{20}$, this means that the distance between any two points on the model (from bow to stern, for example) is one-twentieth the corresponding distance on the real boat. Congruent objects are similar with a scale of 1.

Exercises 16.4

1 (*a*) Before going ahead with construction of a large new type of cargo ship which will be 320 meters long, a scale model is built and tested for seaworthiness in a wave tank. The model is to be built on a scale of $\frac{1}{80}$. How long will the model ship be?

(*b*) How wide is the ship supposed to be if the scale model is to be 0.45 meter wide?

2 (*a*) Many aeronautical charts are printed on a scale of 1:500,000. What does this mean?

(*b*) On a 1:500,000 chart two airports are shown 23 centimeters apart. How many kilometers apart are they in reality?

(*c*) A 1:500,000 chart shows airports at Anacortes, Washington, and Everett, Washington, $5\frac{5}{8}$ inches apart. How many statute miles are these airports apart? (There are 12 inches in a foot, 5,280 feet to a mile.)

3 On a map of West Africa, Abidjan and Accra are $4\frac{3}{4}$ inches apart. The map scale is 1:4,000,000. How many air miles is it from Accra to Abidjan?

4 Draw a triangle and connect the midpoints of its sides by line segments.

(*a*) How are the four resulting small triangles related to each other?

(*b*) How are each of these triangles related to the original triangle?

5 The discovery exercise in Sec. 14.1 involves two pairs of similar triangles.

(*a*) Which ones are they?

(*b*) Since the shape of a triangle is determined by its angles, similar triangles are just those whose corresponding angles are equal. Check this by measuring the angles of the two pairs of similar triangles you found in part (*a*).

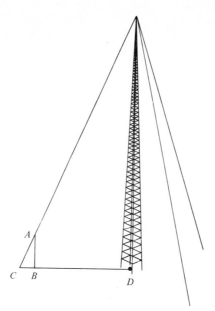

Figure 16.15

6 A slender transmitting tower stands on level ground and is held upright by guy cables, as pictured in Fig. 16.15. To measure the height of the tower, a plumb line is hung from point A and touches the ground at B. The plumb line is 3 meters long, and distances CB and CD are measured as 0.8 and 11 meters, respectively. How tall is the tower?

7 (a) On a sunny day a flagpole casts a 9-meter shadow on level ground, while a meterstick held upright casts a 0.4-meter shadow. How tall is the flagpole?

(b) Could the same method be used on a uniformly sloping hillside?

8 A photo shop sells 5 by 7 prints for $1.75. At that price per square inch how much would an 8 × 10 print cost?

9 (a) In Fig. 16.16, using triangle A as the original, what are the scales of enlargements B, C, and D?

(b) In terms of the area of A, what are the areas of triangles B, C, and D?

(c) If a triangle were made similar to A but on a scale of 10:1, what would its area be in terms of the area of A?

(d) In general, if two triangles are similar and the scale is x, what is the ratio of their areas?

10 Does the result of Prob. 9 apply to rectangles as well? Experiment with the ones shown in Fig. 16.17.

Figure 16.16

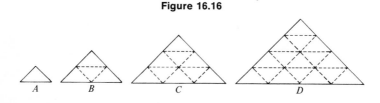

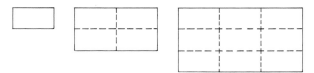

Figure 16.17

11   If the radius of a circle is r units, its area is πr^2 square units. All circles are similar.
   (a) If you multiply the radius of a circle by 2, what do you do to its area?
   (b) If you multiply the radius of a circle by 7, what do you do to its area?
   (c) If you multiply the radius by y, what do you do to the area?
12   All cubes are similar. If a cube 1 unit on edge is taken as a unit of volume. What is the volume of a cube:
   (a) 2 units on an edge   (b) 3 units on an edge
   (c) 4 units on an edge   (d) n units on an edge
13   Problems 9–12 have important practical consequences. They show, for example, why it is not possible to design a supersonic transport plane by simply scaling up a supersonic fighter. Suppose, for example, that an airplane was built by the same plans as a supersonic fighter only 10 times as large.
   (a) The wing area (which determines its lifting power) would be multiplied by ____.
   (b) What would the volume (and therefore the weight) of the airplane be multiplied by?
   (c) Can you see why the airplane would never fly?

CHAPTER 17
POLYGONS AND POLYHEDRA

Here you will meet shapes that have been studied since ancient times for their beauty. More recently they have found use in applications such as crystallography and geodesic domes. One of the best ways to study these figures is to make them yourself. This chapter helps you do so. Building models gives copious practice in the use of ruler and protractor, develops concepts of spatial relations and symmetry, and combines art with mathematics. Models make attractive mobiles, greeting cards, and Christmas tree ornaments. Making models is so much fun that people are often unaware of how much mathematics they are learning as they do it.

17.1 REGULAR POLYGONS

Some of the *polygons* (many-sided figures) in Fig. 17.1 have the symmetry property of looking the same regardless of which side they rest on. These are called *regular* polygons. Which ones are they? Can a polygon have all sides the same length (such a polygon is called *equilateral*) and still not be regular? At first you might not think it could, but there are three such polygons in Fig. 17.1. Can a polygon have all its angles equal and still not be regular? You can find some of these, too, in Fig. 17.1. Evidently, for a polygon to be regular all angles must be equal and all sides must be the same length.

No introduction to polygons is complete without a little terminology. The line segments bounding a polygon are called *edges*, and the corners of the polygon are called *vertices* (singular, *vertex*). The root *poly-* is Greek for "many," but when we specify a polygon with precisely *n* sides we call it an *n-gon*. For example, one with 93 sides is a 93-gon. Polygons with relatively few sides have special names; those with five, six, seven, eight, and ten sides are called *pentagons, hexagons, heptagons, octagons,* and *decagons,* respectively. If the names followed a consistent pattern, polygons with three and

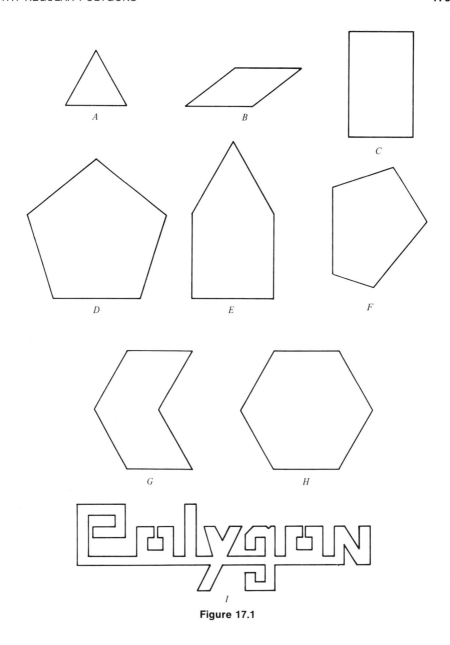

Figure 17.1

four sides would be called trigons and tetragons, but they are called more prosaically *triangles* and *quadrilaterals*.

For work in subsequent sections you will need to know how to make reasonably accurate regular polygons. The following exercises will help with that.

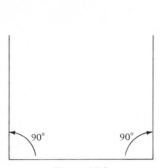

Figure 17.2   **Figure 17.3**

Exercises 17.1

1 (*a*) If the angles in any triangle total 180°, how large is each angle in a regular triangle?
   (*b*) Make a regular triangle by first making a side, then making the appropriate angles at the
   ends of that side, and finally extending the lines thus determined until they meet.
   (*c*) A polygon is *equilateral* if its sides are all the same length. Figure 17.1 shows several
   equilateral polygons which are not regular. Is every equilateral triangle regular?
2 Here is a way to construct a square.
   (*a*) Make one side (about 2 inches long is a convenient size).
   (*b*) At each end make a 90° angle, as shown in Fig. 17.2.
   (*c*) Extend the two new sides to make them as long as the first.
   (*d*) Connect the ends of these two new sides to finish the square.
   (*e*) Check your square for symmetry by looking at it from various directions.
   (*f*) Is every equilateral quadrilateral a square? Justify your answer.
3 Here is a way to make a regular pentagon.
   (*a*) First determine the size of the angles. (For this you may want to review Probs. 14 to 17 in
   Exercises 16.2 before going on.) Imagine an ant who walks once around the border of a regular
   pentagon (Fig. 17.3), starting and ending the trip at point *A*. During the trip he makes a turn
   through an angle *x* at each vertex as shown, and when he gets back to *A* he is facing in the
   same direction as when he set out. How large is each angle *x*?
   (*b*) Build a regular pentagon 4 inches on a side as follows:
       (1) Make the first side.
       (2) At each end of the first side make the appropriate angle.
       (3) Use the angles you made to determine the sides 4 inches long adjoining the one you
       began with.

Figure 17.4

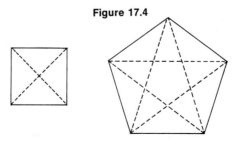

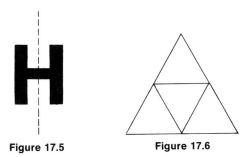

Figure 17.5   Figure 17.6

(4) Make the appropriate angles at the ends of the sides formed in step 3.

(5) Draw the sides determined by the angles in step 4 to complete the pentagon.

(c) Inspect your pentagon for symmetry. If it is a good job, save it for later work (Sec. 17.2).

4 (a) Make a regular hexagon, using the same general method as in the previous examples.

(b) Draw the lines connecting the three pairs of opposite vertices (corners) of the hexagon you made in part (a). What do you notice about the triangles formed?

5 A *diagonal* of a regular polygon is a line segment connecting vertices which are not ends of the same side. For example, the diagonals are dashed in the square and pentagon in Fig. 17.4.

(a) Fill in the table below and find the rule in it.

Number of sides	3	4	5	6	7	8
Number of diagonals	0	2	5			

Hint: $3 \cdot 0 = ?, 4 \cdot 1 = ?, 5 \cdot 2 = ?, 6 \cdot 3 = ?, \ldots$

(b) How many diagonals has a regular polygon with 100 sides?

(c) Can you explain the rule in part (a)? *Hint*: How many diagonals meet at each vertex?

(d) How is this related to Prob. 6 of Exercises 3.1?

6 (a) If the paper is folded along the dashed line, the two halves of the H in Fig. 17.5 will match. The dashed line is therefore known as a *line of symmetry* for this figure. Can you find another?

(b) The figure in Fig. 17.6 has three lines of symmetry. Find them.

7 (a) Draw in the lines of symmetry of the quadrilaterals in Fig. 17.7.

(b) What kind of quadrilateral has the most symmetry?

(c) In general, what kind of *n*-gon do you think has the most symmetry?

Figure 17.7

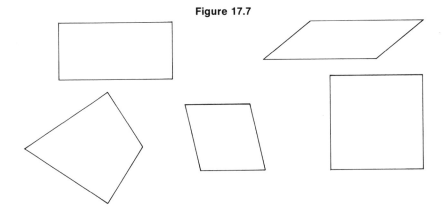

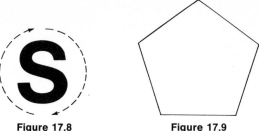

Figure 17.8   **Figure 17.9**

8 The letter S has no lines of symmetry, but it does have what is called *rotational symmetry* in that it looks the same after a rotation of one-half turn about its center point (Fig. 17.8). More specifically, this figure is said to have 180° rotational symmetry, because a rotation of 180° is the smallest that will bring it into what looks like its original position.

    (*a*) Which polygons in Fig. 17.7 have 180° rotational symmetry?

    (*b*) Figure 17.6 does not have 180° rotational symmetry, but it does have 120° rotational symmetry about its center point. How is that point related to the lines of symmetry?

9 (*a*) Find the lines of symmetry of the regular pentagon in Fig. 17.9.

    (*b*) This figure also has ____ (how many degrees?) rotational symmetry.

    (*c*) How is the center of rotational symmetry related to the lines of symmetry?

10 Could the three lines in Fig. 17.10 all be lines of symmetry of the same polygon? Justify your answer in terms of observations made in earlier problems.

11 (*a*) Find three figures in Figs. 17.5–17.7 with just two lines of symmetry each. At what angles do the lines of symmetry meet?

    (*b*) Do these three figures also have rotational symmetry? If so, through how many degrees?

12 (*a*) Figure 17.6 has just three lines of symmetry. Does it also have rotational symmetry? If so, through how many degrees?

    (*b*) Find a figure in Fig. 17.7 with just four lines of symmetry. Does it also have rotational symmetry? If so, through how many degrees?

    (*c*) From Probs. 11 and 12, would you expect a figure with just 10 lines of symmetry to have rotational symmetry? If so, through how many degrees?

13 Sketch figures with 120, 90, and 72° rotational symmetry but no lines of symmetry. *Hint*: Think of propeller blades.

17.2 TESSELLATIONS

Tiled floors are usually made with square tiles as in Fig. 17.11*a*, but you may also have seen hexagonal and triangular tilings, or *tessellations*, as they are called, like those in Fig. 17.11*b* and *c*.

Figure 17.10

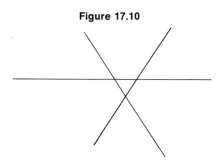

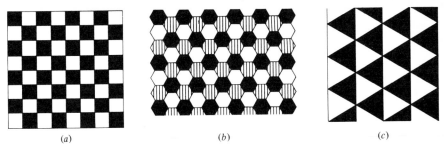

Figure 17.11   The regular tessellations

These are the only ways to fill a plane with congruent regular polygons, at least if we do not allow vertices of some polygons to lie along edges of others, as in Fig. 17.12. The tessellations in Fig. 17.11 are called *regular* tessellations, because each uses congruent copies of only one regular polygon. In the exercises below you will see why there are no other regular tessellations.

There are more possibilities if we are willing to use two or more kinds of regular polygons simultaneously. For example, if the squares in Fig. 17.11*a* are *truncated*, i.e., their corners are cut off, to produce regular octagons, as in Fig. 17.13, we get the tessellation in Fig. 17.14. Similarly, truncating the hexagons in Fig. 17.11*b* yields the tessellation in Fig. 17.15. If the triangles in Fig. 17.8*c* are truncated to produce hexagons, as in Fig. 17.16, the result will be the tessellation in Fig. 17.11*c*. If the truncation is continued further to produce the inner triangle in Fig. 17.16, the effect on the tessellation in Fig. 17.11*c* is to transform it into the one in Fig. 17.17.

The tessellations in Fig. 17.14, 17.15, and 17.17 are called *semiregular*, since although they use more than one kind of polygon, those used are all regular and each vertex is exactly like every other. In fact, semiregular tessellations are sometimes described by describing the kinds of regular polygons one would meet walking around a vertex. In Fig. 17.14 we might start with the square, then encounter the two octagons, so that tessellation has the *vertex symbol* (4,8,8). Similarly, vertex symbols for the tessellations in Figs. 17.15 and 17.17 are (3,12,12) and (3,6,3,6), respectively.

Figure 17.13   Truncating a square

Figure 17.12

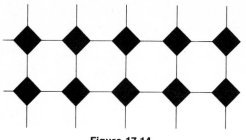

Figure 17.14

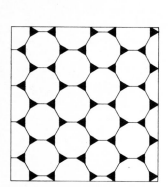

Figure 17.15

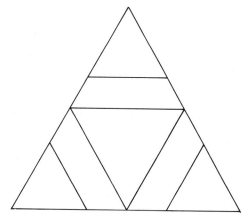

Figure 17.16

Figure 17.17

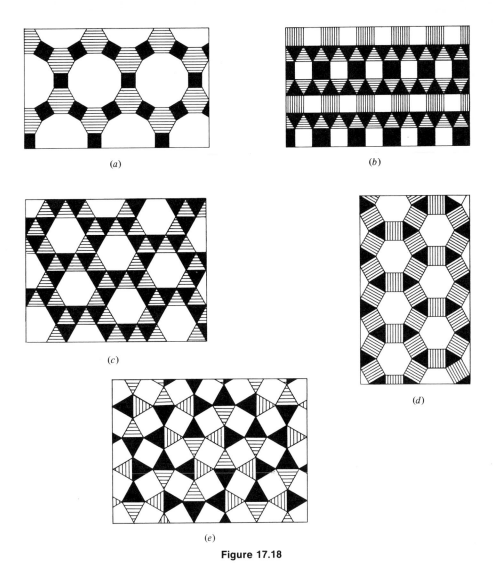

Figure 17.18

Exercises 17.2

1 There are only eight semiregular tessellations in all, three of which are pictured in Figs. 17.14, 17.15, and 17.17. The other five are shown in Fig. 17.18. What are their vertex symbols?

2 (a) Suppose we add the angles that meet at a given vertex of a tessellation. For the regular tessellations we have (see Fig. 17.11):

(1) Four squares, $90° + 90° + 90° + 90° =$

(2) Three hexagons, $120° + 120° + 120° =$

(3) Six triangles, $60° + 60° + 60° + 60° + 60° + 60° =$

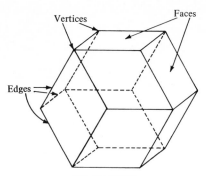

Vertices

Faces

Edges

Figure 17.19

In each case what is the total? Do you see why this must be?

(*b*) Are there any regular tessellations other than those shown in Fig. 17.11? Justify your answer with evidence from part (*a*).

3   The interior angle of a regular 12-gon is 150°. Therefore, since the angles at a vertex add to 360°, one might expect to be able to make semiregular tessellations with vertex symbols (3,3,4,12) and (3,4,3,12). Try it and see for yourself what goes wrong.

17.3 REGULAR POLYHEDRA

The three-dimensional analogs of polygons and tessellations are called *poly-hedra* (singular, *polyhedron*). The polygonal parts of planes which bound a

Figure 17.20

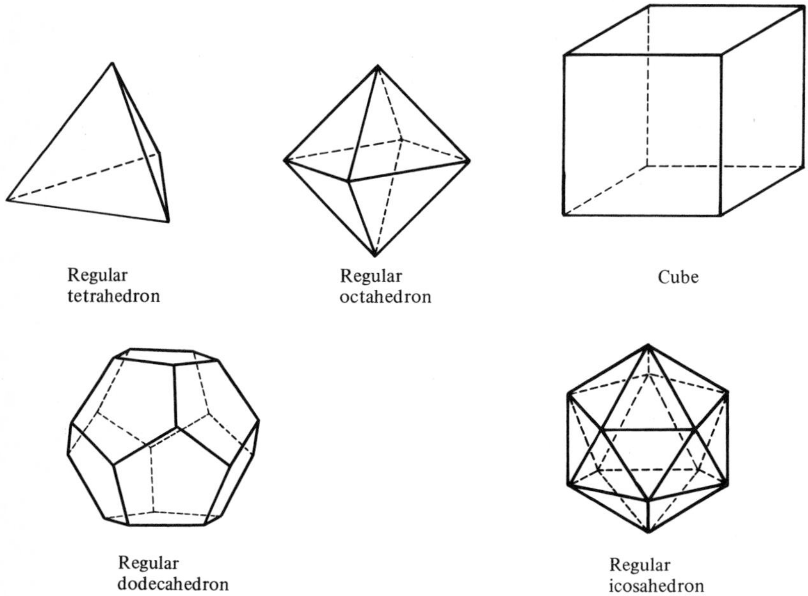

Regular
tetrahedron

Regular
octahedron

Cube

Regular
dodecahedron

Regular
icosahedron

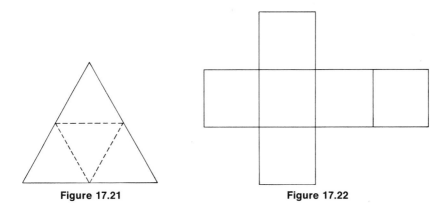

Figure 17.21   Figure 17.22

polyhedron are called its *faces*, the lines where adjoining faces meet are called *edges*, and the points where adjoining edges meet are called *vertices*. The polyhedron illustrating these terms in Fig. 17.19 is a common crystal form of the mineral garnet.

A polyhderon is *regular* if its faces are identical regular polygons, the same number of edges meet at each vertex, and all angles between adjoining faces are equal. As Lewis Carroll aptly said, there are "distressingly few" regular polyhedra. In fact the only kinds are those in Fig. 17.20. Of these five, the regular tetrahedron is the easiest to build. One way is to connect the midpoints of the sides of an equilateral triangle, as in Fig. 17.21, fold along these lines, and tape the result together. You may be familiar with cream containers in this shape. The regular tetrahedron is also the crystal form of a metallic mineral, known appropriately as tetrahedrite.

A model cube is also easy to make, either by taping together six squares or by folding and taping a pattern like the one in Fig. 17.22. Figures 17.23 and 17.24 show similar patterns for the regular octahedron and icosahedron. Many

Figure 17.23

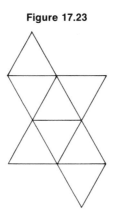

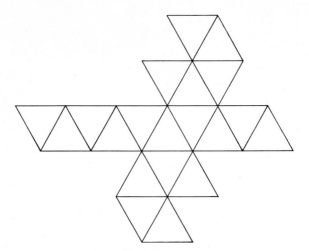

Figure 17.24

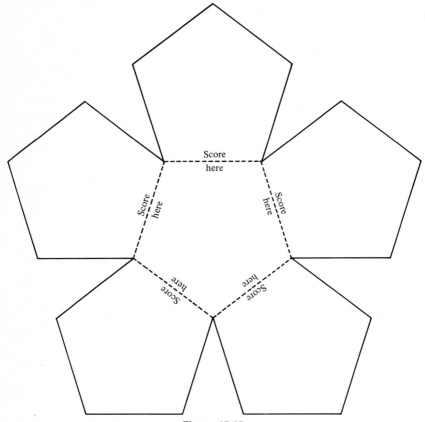

Score
here

Score
here

Score
here

Score
here

Score
here

Figure 17.25

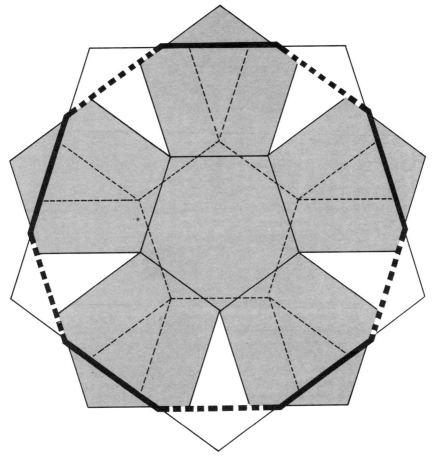

Figure 17.26

minerals, including table salt, occur naturally in cubical crystals, and others, including diamonds, are found in octahedral crystals, but there are no minerals at all whose crystals are regular icosahedra or regular dodecahedra.

Here is an enjoyable construction for the regular dodecahedron. Make two identical figures like Fig. 17.25 out of sturdy cardboard (the kind that comes with pads of paper). The pieces should be scored (cut part way through) along the lines where pentagons meet so that they flex along these lines. When an elastic band is woven about the two pieces, as in Fig. 17.26, they spring up into a regular dodecahedron, as in Fig. 17.27. It would be tedious to make 12 separate regular pentagons, but fortunately there is a simple way to build the shape in Fig. 17.25 from a single large regular pentagon:

1. Make a large regular pentagon about 4 inches on a side.

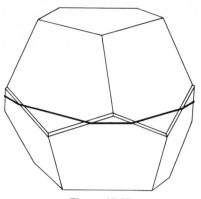

Figure 17.27

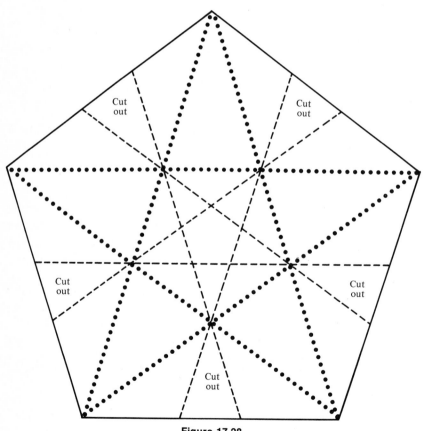

Cut
out

Cut
out

Cut
out

Cut
out

Cut
out

Figure 17.28

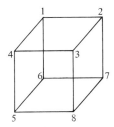

Figure 17.29

2. Draw the diagonals of this pentagon (dotted in Fig. 17.28). This forms a star with a small regular pentagon inside it.

3. Draw the diagonals of this small pentagon and extend them out to the edges of the large pentagon. (These are dashed lines in Fig. 17.28).

4. Cut out the small triangles formed in step 3, as indicated in Fig. 17.28.

Exercises 17.3

1 In the cube in Fig. 17.29 what figure would be formed by line segments connecting the odd-numbered vertices?

2 In the dodecahedron in Fig. 17.30 eight of the vertices are marked with heavy dots. Of what figure are they the vertices? (You may have trouble seeing this unless you use a model.)

3 Show how to select four vertices of a regular dodecahedron so that these vertices are also vertices of a regular tetrahedron. *Hint*: Use the results of Probs. 1 and 2.

4 Describe the figure whose vertices are the midpoints of the faces of a
   (*a*) Regular tetrahedron   (*b*) Cube
   (*c*) Regular octahedron   (*d*) Regular dodecahedron
   (*e*) Regular icosahedron

*5 Describe how to slice a cube to get a cross section which is a regular hexagon.

6 In a tessellation the angles at each vertex add up to 360°. In a regular polyhedron, however, the

Figure 17.30

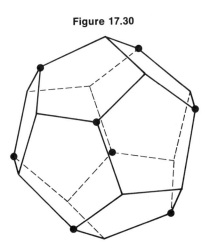

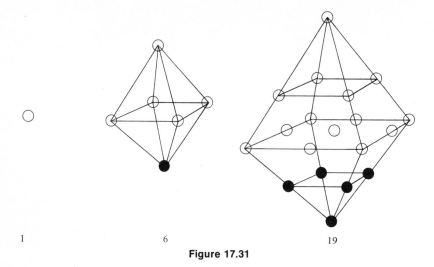

Figure 17.31

angles at each vertex add up to less than 360°. Together with the fact that each vertex of a polyhedron is the junction of at least three faces, this can be used to show why there are only five regular polyhedra.

(*a*) The tetrahedron, octahedron, and icosahedron have three, four, and five equilateral triangles meeting at each vertex. Why can't there be a regular polyhedron with six or more triangles meeting at each vertex?

(*b*) The cube has three squares meeting at each vertex. Why can't there be a regular polyhedron with four or more squares meeting at each vertex?

(*c*) The dodecahedron has three regular pentagons at each vertex. Why can't there be a regular polyhedron with four or more regular pentagons meeting at each vertex?

(*d*) Why can't there be a regular polyhedron built from polygons with six or more sides?

7 What figure is formed by connecting the midpoints of the edges of a regular tetrahedron?

8 Figure 3.13 shows how two consecutive tetrahedral numbers can be put together to get a number that may be associated with balls stacked in a four-sided pyramid. Putting two of these in a row (indicated by shaded and unshaded balls in Fig. 17.31) together yields what are called *octahedral* numbers, 1, 6, 19, ... In the light of Prob. 9 in Exercises 3.2, these can be found by adding two consecutive pairs of consecutive tetrahedral numbers; for example,

$$(1 + 4) + (4 + 10) = 19$$

$$(4 + 10) + (10 + 20) = 44$$

$$(10 + 20) + (20 + 35) = 75$$

and so on are octahedral numbers.

(*a*) Find the next three octahedral numbers.

(*b*) Problem 7 suggests that if four tetrahedral numbers of a suitable size are added to an octahedral number, the result will be a larger tetrahedral number. With this in mind, compute

$$1 + 4 \cdot 0 =$$

$$6 + 4 \cdot 1 =$$

$$19 + 4 \cdot 4 =$$

$$44 + 4 \cdot 10 =$$

(*c*) How does this relate to Prob. 13 of Exercises 3.2?

(*d*) Generalizing, Fig. 17.32 suggests that 4 times an octahedral number and 10 times a tetrahedral number add to a larger tetrahedral number. This is not quite so, but almost. Compute

$$4 \cdot 1 + 10 \cdot 0 + 0 =$$
$$4 \cdot 6 + 10 \cdot 1 + 1 =$$
$$4 \cdot 19 + 10 \cdot 4 + 4 =$$
$$4 \cdot 44 + 10 \cdot 10 + 10 =$$

What are the results? Extend the pattern another line or two and check it.

17.4 SEMIREGULAR POLYHEDRA

Relaxing the strict requirements of regularity, we now turn to semiregular polyhedra. Like semiregular tessellations, semiregular polyhedra are made with at least two kinds of regular polygons and have identical sets of polygons meeting at every vertex.

You will get more out of this section if you actually build the polyhedra. The illustrations, together with the vertex symbols and numbers of each kind of polygon needed, give you all the necessary information. Once you have made the polygons to be used as faces, tape some together to form a vertex as indicated by the vertex symbol and figure. Then you will see what pieces to add to form another vertex, and as you continue, the model will begin to resemble the illustration.

Do not be dismayed at the large numbers of polygons some of these shapes require. Equilateral triangles can be produced six at a time, as in part (*b*) of Prob. 4 of Exercises 17.1, and pentagons can be mass-produced by the method outlined in the discussion of the regular dodecahedron. This is particularly

Figure 17.32

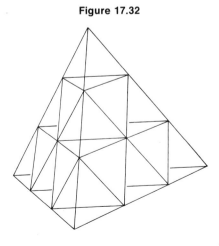

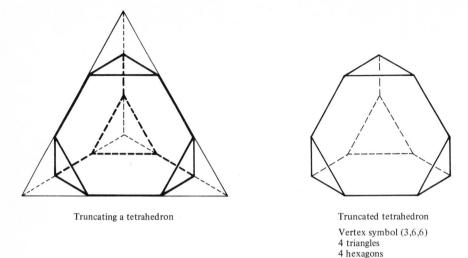

Truncating a tetrahedron

Truncated tetrahedron
Vertex symbol (3,6,6)
4 triangles
4 hexagons

Figure 17.33

convenient because, except for the 5-prism and the alternating 5-prism, any semiregular polyhedron which has pentagonal faces has exactly 12 of them.

Do not hurry your model building. Craftsmanship and precision are important; so is the choice of colors. Start with the simpler models; in the

Figure 17.34

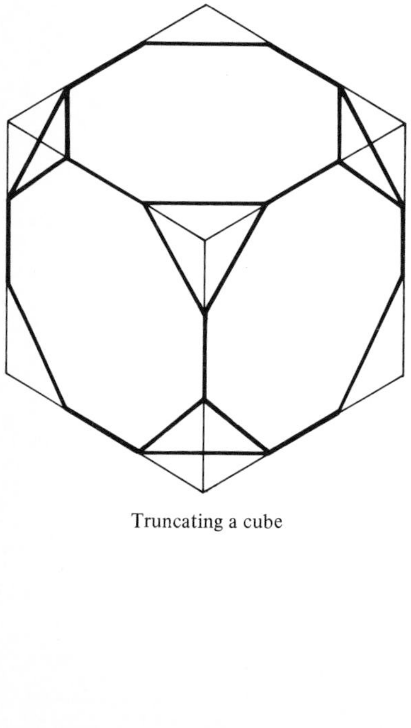

Truncating a cube

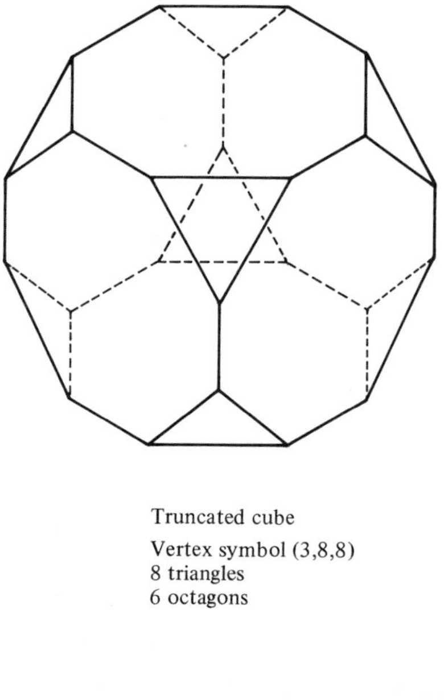

Truncated cube
Vertex symbol (3,8,8)
8 triangles
6 octagons

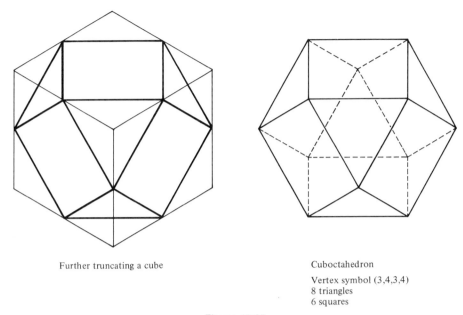

Further truncating a cube

Cuboctahedron

Vertex symbol (3,4,3,4)
8 triangles
6 squares

Figure 17.35

beginning one of these is plenty for a whole evening's work. As you gain experience, you will refine your techniques and be able to do even some of the more complex models in an evening. *Caution*: do not try to make many models at once; an overdose of this otherwise pleasant work can make it tedious.

Many of these polyhedra may be obtained by cutting the vertices off of other polyhedra, that is, *truncating* them. Two more pairs of semiregular polyhedra may be obtained by modifying the truncation process. Figure 17.37, for example, shows that truncating a cuboctahedron yields something like a great rhombicuboctahedron but with squares replaced by long, thin rectangles. Compressing the truncated cuboctahedron produces a great rhombicuboctahedron. Similar remarks apply to the rhombicuboctahedron (Fig. 17.38), the great rhombicosidodecohedron (Fig. 17.43), and the rhombicosidodecahedron (Fig. 17.44). Look at a rhombicuboctahedron or its illustration (Fig. 17.38). Some squares adjoin triangles, and others do not. If all the squares which adjoin triangles are replaced by a pair of triangles, a slight twist is introduced and the result is a snub cube (Fig. 17.39). Applying the same process to a rhombicosidodecahedron produces a snub dodecahedron (Fig. 17.45). The snub dodecahedron and snub cube come in two different forms, each the mirror image of the other. The forms differ in the direction of the twist introduced by replacing squares with triangles.

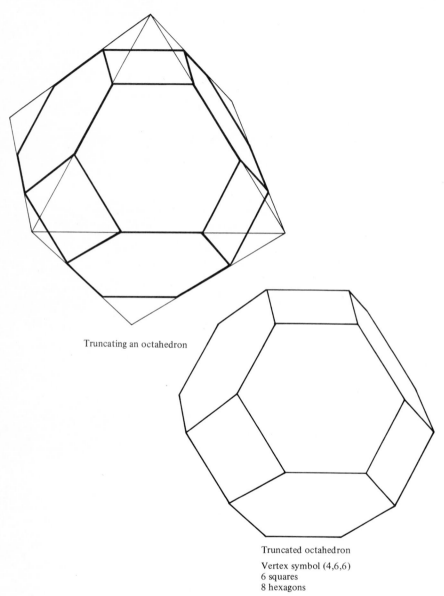

Truncating an octahedron

Truncated octahedron

Vertex symbol (4,6,6)
6 squares
8 hexagons

Figure 17.36

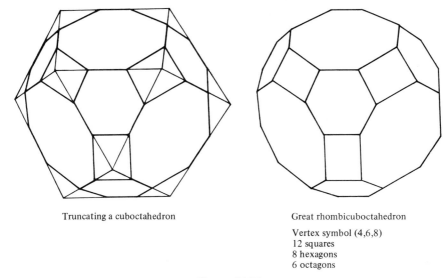

Truncating a cuboctahedron

Great rhombicuboctahedron

Vertex symbol (4,6,8)
12 squares
8 hexagons
6 octagons

Figure 17.37

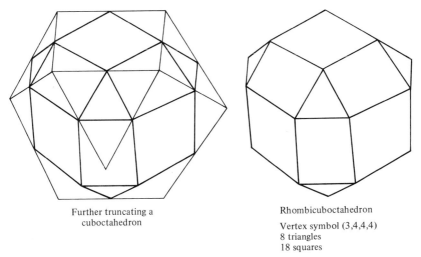

Further truncating a
cuboctahedron

Rhombicuboctahedron

Vertex symbol (3,4,4,4)
8 triangles
18 squares

Figure 17.38

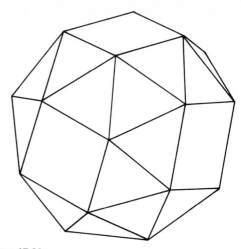

Figure 17.39   Snub cube

Vertex symbol (3,3,3,3,4)
32 triangles
6 squares

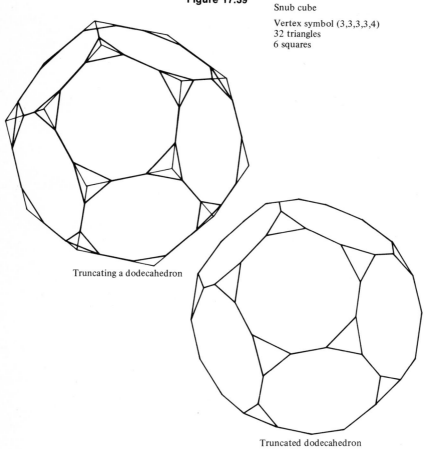

Truncating a dodecahedron

Truncated dodecahedron

Figure 17.40   Vertex symbol (3,10,10)
20 triangles
12 decagons

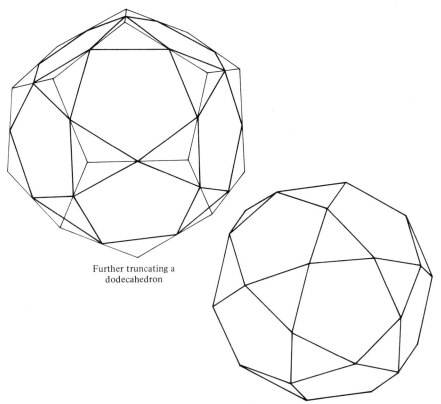

Further truncating a
dodecahedron

Figure 17.41   Icosidodecahedron

Vertex symbol (3,5,3,5)
20 triangles
12 pentagons

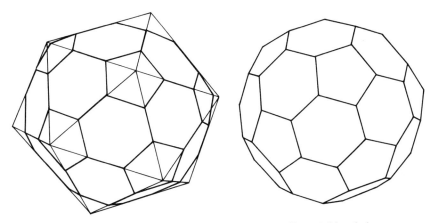

Truncating an icosahedron

Truncated icosahedron

Figure 17.42   Vertex symbol (5,6,6)
12 pentagons
20 hexagons

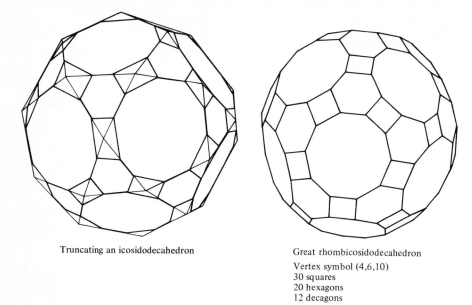

Truncating an icosidodecahedron

Great rhombicosidodecahedron

Vertex symbol (4,6,10)
30 squares
20 hexagons
12 decagons

Figure 17.43

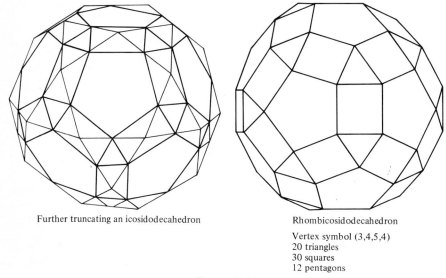

Further truncating an icosidodecahedron

Rhombicosidodecahedron

Vertex symbol (3,4,5,4)
20 triangles
30 squares
12 pentagons

Figure 17.44

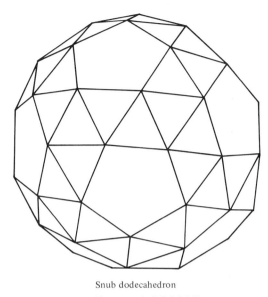

Snub dodecahedron

Vertex symbol (3,3,3,3,5)
80 triangles
12 pentagons

Figure 17.45

In addition to the semiregular polyhedra mentioned above there are two other infinitely large families of semiregular polyhedra. One is the *n*-prisms, shown in Fig. 17.46 for *n* = 3 and 5. (What is the common name for the 4-prism?) The other is the alternating prisms, shown in Fig. 17.47 for *n* = 6. (What is another name for the alternating 3-prism?)

There are many other interesting polyhedra, some even more spectacular than those considered here. If you let the faces of a polyhedron intersect each other, then the *great dodecahedron* (Fig. 17.49) and great icosahedron (Fig. 17.52) must be considered regular polyhedra. If you also let regular star polygons like the one in Fig. 17.48 be used for faces, then the polyhedra shown in Figs. 17.50 and 17.51 must also be considered regular. Information on building these and others can be found in the literature (see the references by Cundy and Rollett, Holden, and Wenninger in the Bibliography).

Figure 17.46

3-prism   5-prism

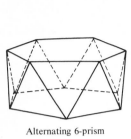

Alternating 6-prism

Figure 17.47

Figure 17.48

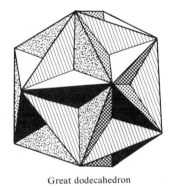

Great dodecahedron

Figure 17.49

Great stellated dodecahedron

Figure 17.50

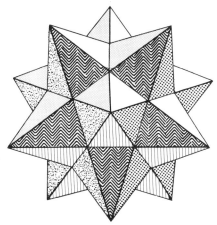

Small stellated dodecahedron
Figure 17.51

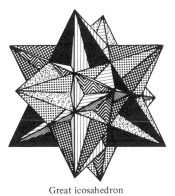

Great icosahedron
Figure 17.52

CHAPTER 18
TOPOLOGICAL QUESTIONS

We now, turn to some geometric questions which do not involve lengths or angles. In this chapter you will use numbers to count but not to measure. Despite this limitation, these questions stimulated the development of a whole branch of mathematics called *topology*. The main point of this chapter, however, is to show you again that you can discover mathematics for yourself, even where you might not have thought of looking.

18.1 A PATTERN

Are the numbers of faces, vertices, and edges of a polyhedron related? No doubt you realize they must be, for otherwise the question would not have been raised here. You are right. Indeed raising the question is the crucial step, for the pattern is easily discovered by anyone who bothers to look for it. Here you will rediscover it for yourself, and you will see again that skill at asking questions can be as valuable in mathematics as skill at answering them.

As usual, a good way to begin looking for a pattern is to gather some information. Count the faces, vertices, and edges of some polyhedra. Then organize the data and look for a pattern.

In this case the main problem will be to do the counting accurately. Fortunately, we can get plenty of data from the regular and semiregular polyhedra, for which the counting can be done systematically. The following examples illustrate the method.

REGULAR DODECAHEDRON

We saw in Chap. 17 that this has 12 regular pentagons as faces. To count edges, first imagine that the 12 pentagons have been made but not yet put together. Each pentagon has 5 edges, so the 12 pentagons have a total of 60 edges. When the pentagons are put together, two of their edges will form each edge of the dodecahedron, so it has $60 \div 2 = 30$ edges. Similarly, the unassembled pentagons have $5 \cdot 12 = 60$ vertices, and since (see Fig. 17.20) 3 of these go together to form each vertex of the dodecahedron, it has $60 \div 3 = 20$ vertices.

SNUB CUBE

We saw in Chap. 17 that a snub cube is made with 6 squares and 32 triangles, and so it has 38 faces in all. Unassembled, the squares and triangles have a total of $6 \cdot 4 + 32 \cdot 3 = 120$ edges. When the pieces are put together, each edge of the snub cube will be formed from 2 edges of the separate polygons, so the snub cube has $120 \div 2 = 60$ edges. Similarly, the unassembled polygons have $6 \cdot 4 + 32 \cdot 3 = 120$ vertices in all, and since 5 of them are needed to form each vertex of the snub cube (see Fig. 17.39) it has $120 \div 5 = 24$ vertices.

At this point you may feel you understand the counting method. If you do, skip directly to Prob. 3 of the following exercises. Otherwise try Probs. 1 and 2 first.

Exercises 18.1

1   A regular icosahedron is made from 20 triangles.
   (*a*) Unassembled, these triangles have _____ edges in all.
   (*b*) When the triangles are put together, _____ edges are used to make each edge of the icosahedron.
   (*c*) The icosahedron has _____ ÷ _____ = _____ edges.
   (*d*) The unassembled triangles have _____ vertices in all.
   (*e*) How many triangles meet at each vertex of the icosahedron? *Hint*: see Fig. 17.20.
   (*f*) The icosahedron has a total of _____ ÷ _____ = _____ vertices.
2   A truncated tetrahedron is made from 4 hexagons and 4 triangles.
   (*a*) Unassembled, these pieces have a total of _____ edges.
   (*b*) When the pieces are put together, how many of their edges go to make each edge of the polyhedron?
   (*c*) The truncated tetrahedron has _____ ÷ _____ = _____ edges.
   (*d*) The unassembled polygons have _____ · 6 + _____ · 3 = _____ vertices in all.
   (*e*) How many faces meet at each vertex of the truncated tetrahedron?
   (*f*) The truncated tetrahedron has _____ ÷ _____ = _____ vertices.
3   (*a*) Organize your data by filling in the rest of this table.

Polyhedron	Number of faces = F	Number of vertices = V	Number of edges = F
Tetrahedron			
Cube			
Octahedron			
Dodecahedron	12	20	30
Icosahedron			
Truncated tetrahedron			
Truncated cube			
Snub cube	38	24	60

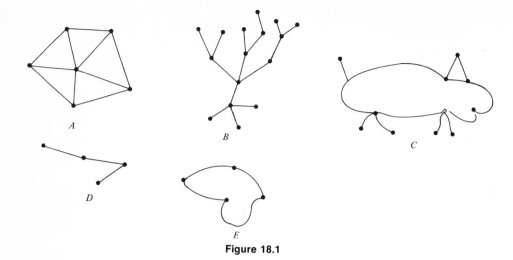

Figure 18.1

(*b*) Look for a pattern in the table. *Hint*: Add 2 to the number of edges in each case. What is the result?

(*c*) Test your pattern on other polyhedra, such as prisms and pyramids. Does it hold in all cases you tried?

(*d*) Can you think of how this discovery might be generalized?

18.2 NETWORKS

Whenever you make a discovery, try to generalize it. This sometimes leads to new discoveries, and it often brings out what is essential in what you found to begin with. In the previous section our discovery was ostensibly about the numbers of faces, vertices, and edges of polyhedra. Of polyhedra? Did we ever use the flatness of the faces or the straightness of the edges? Would the relationship still hold for a polyhedron curved and bent a bit? If our discovery can be generalized to deformed polyhedra, then it is not fundamentally about polyhedra at all. It is about something more general called a *network*.

A *network* is a collection of points (called *vertices*) connected by segments of lines or curves (called *edges*). Figure 18.1 shows some networks.

These networks all lie in a plane, while the polyhedra do not, but with each polyhedron we can associate a planar network by projecting it onto a plane, as shown for a tetrahedron in Fig. 18.2. You will find projections of some other polyhedra in the exercises below. In such a projection the vertices of the polyhedron correspond to those of the network, and the edges of the polyhedron correspond to those of the network.

All but one of the faces of the polyhedron correspond to regions enclosed by edges of the network. If we think of the extra face as corresponding to the region of the plane outside of the network, then it is reasonable to expect that the pattern you found for faces, vertices, and edges of polyhedra carries over to networks.

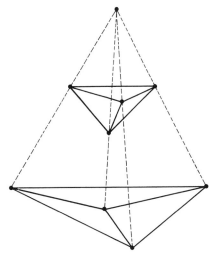

Figure 18.2

Exercises 18.2

1 Each network in Fig. 18.3 is a projection of a regular or semiregular polyhedron. Can you name which?

2 (*a*) For this exercise, refer to the networks in Fig. 18.1. Count the number of regions the plane is divided into by the network (do not forget the outside), the number of vertices, and the number of edges. Summarize your findings in a table like this:

Figure 18.3

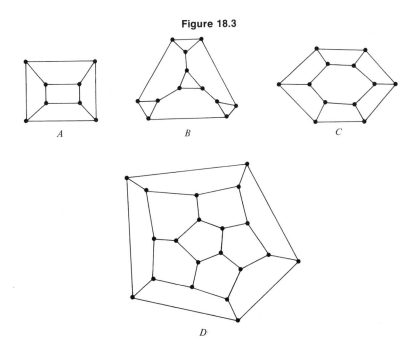

A   B   C

D

Network	Number of regions = R	Number of vertices = V	Number of edges = E
A			
B			
C			
D			

Does the pattern from Sec. 18.1 still hold?

(b) To see what is happening, consider how a network is built up from the simplest possible network, which is a single vertex. For this simple network there are no edges, and only one region, so

$$R + V - E = 1 + 1 - 0 = 2$$

Now suppose we draw an edge from this vertex. If the edge loops around and ends at the same vertex, like this $\times\!\circ$, then we have not added a new vertex but we have enclosed a region. If the edge does not loop, it does not enclose a face, but it adds a new vertex, like this $\frown\!$. In the first case $R + V - E$ becomes $2 + 1 - 1 = 2$, and in the second case it is $1 + 2 - 1 = 2$. Similarly, as we draw a network, each time we add an edge we must either add a new vertex or end it at a vertex we have already drawn, thus adding a face. Therefore increasing E by 1 involves increasing either R or V but not both by 1. Thus $R + V - E$ remains constant.

3 (a) Count the faces, vertices, and edges of the polyhedron with a hole in it in Fig. 18.4.
   (b) Does the pattern you found above hold here?
   *(c) Sketch or make other polyhedra with holes in them.
   *(d) Try to modify the pattern you found above to account for the number of holes in tunneled polyhedra.

4 The *degree* of a vertex in a network is the number of edges that end there.
   (a) Find the degrees of the vertices in these networks:

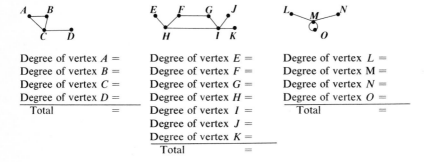

Degree of vertex A =	Degree of vertex E =	Degree of vertex L =
Degree of vertex B =	Degree of vertex F =	Degree of vertex M =
Degree of vertex C =	Degree of vertex G =	Degree of vertex N =
Degree of vertex D =	Degree of vertex H =	Degree of vertex O =
Total =	Degree of vertex I =	Total =
	Degree of vertex J =	
	Degree of vertex K =	
	Total =	

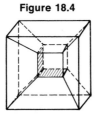

Figure 18.4

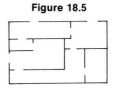

Figure 18.5

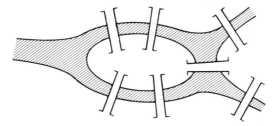

Figure 18.6

(*b*) Count the number of edges in each of the networks above and observe how in each network the number of edges is related to the total of the vertex degrees. Check this pattern on some other networks, such as those in Figs. 18.1 and 18.3.

(*c*) If a network had 50 edges, what do you predict for the sum of its vertex degrees?

(*d*) Can you explain the pattern in part (*b*)? *Hint*: Since the degree of a vertex is the number of edges that end there, the sum of the vertex degrees is the total number of ends of edges in the network. How many ends has each edge?

(*e*) If a network had an odd number of vertices with odd degrees, would the sum of its vertex degrees be even or odd?

(*f*) In light of part (*d*), can the sum of the vertex degrees of a network ever be odd?

(*g*) On the basis of your answers to the above, what can you conclude about the number of odd vertices in a network?

5   A certain polyhedron with no holes in it has 60 vertices with 4 edges meeting at each vertex.

(*a*) How many edges has this polyhedron?

(*b*) How many faces has this polyhedron? [Use your answer from part (*a*), together with the pattern from Sec. 18.1.]

6   The floor plan of a house is shown in Fig. 18.5. Why must any such house have an even number of rooms that have an odd number of doors? *Hint*: How does this relate to Prob. 4?

18.3 SOME PATH QUESTIONS.

Topology was originated as a branch of mathematics with Euler's paper of 1736 on the bridges of Königsberg, an East Prussian town (now Kaliningrad) on the Pregel River.[1] Local tradition said that it was impossible to take a stroll crossing each of the town's seven bridges (Fig. 18.6) just once and ending where you began. Euler proved it. He first observed that the question amounts to asking whether it is possible to trace all the edges of the network in Fig. 18.7 without picking up the pencil in such a way that no edge is traced more than once and you end up where you began. For a path to trace every edge, it must pass through each vertex, and since no edge may be retraced, it must leave each vertex along an edge different from the one on which it arrived at that

[1]Leonhard Euler (1707–1783) was from Basel, Switzerland but spent most of his life in Berlin and St. Petersburg (now Leningrad). Mathematics came so naturally to him that he often worked on problems while playing with some of his 13 children. He was to mathematics roughly what Bach was to music, producing a staggering amount of high-quality work. A collection of his works, still not complete, already fills more than 100 full-sized volumes. Euler's ability and productivity continued unabated even after he had become totally blind.

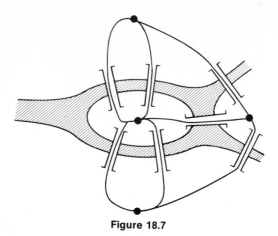

Figure 18.7

vertex. Hence each vertex must be the junction of an even number of edges, or, in the language of Prob. 4 in Exercises 18.2, each vertex must have an even degree. Since all four vertices in Fig. 18.7 have odd degrees, no suitable path can be found. Because of Euler's pioneering work in this field, a path which traces each edge of a network just once and ends where it began is known today as an *Eulerian path*. We have seen that if a network has any odd vertices, it cannot possibly have an Eulerian path.

A more difficult network problem goes back to a puzzle invented in the middle of the nineteenth century by Hamilton.[2] The puzzle was a regular dodecahedron whose vertices were labeled as cities; its object was to find a path along the edges of the dodecahedron (not all edges need be used) which passes through each vertex just once and ends at its starting point. The puzzle is not very difficult, and you can probably solve it yourself in a few minutes (see

[2]William Rowan Hamilton (1805–1865) was born in Dublin and spent most of his life there. By the age of eight he could read Latin, Greek, Hebrew, Italian, and French as well as English, but soon his interests turned toward science. At twenty-one he published a paper on mathematical optics which led to his appointment as professor at Trinity College in Dublin while still an undergraduate there.

Figure 18.8

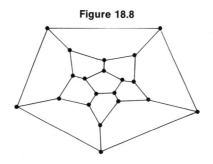

Prob. 3 below). However, the general problem of determining for which networks such a *Hamiltonian path* can be found is considerably more difficult and is not completely solved to this day.

Like so much of mathematics, networks were first regarded as curiosities, but lately they have taken on practical importance. The study of networks is now used in designing electric circuits, flood-control dam systems, telephone exchanges, and pipeline and power grids as well as in analyzing nervous and circulatory systems.

Exercises 18.3

1   For which of these networks can Eulerian paths be found?

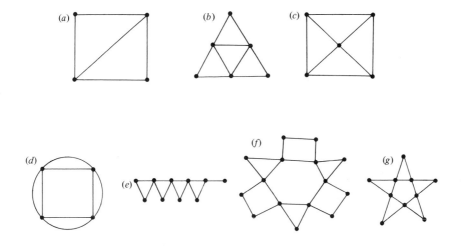

2   (*a*) For which of the networks in Prob. 1 is it possible to find a path which traces each edge just once if we are willing to drop the requirement that the path ends where it began?
   (*b*) For a network to have a path that meets the conditions in part (*a*), how many vertices of odd degree can it have?
3   The network in Fig. 18.8 can be obtained by projecting a regular dodecahedron.
   (*a*) Find a Hamiltonian path in this network.
   (*b*) Does this network have an Eulerian path?
4   (*a*) Is it possible to arrange a drive which crosses each bridge and tunnel in New York City just once and ends where it starts? Use the map in Fig. 18.9, which omits bridges and tunnels not allowing automobiles.
   (*b*) Would the problem be possible if the requirement of starting and ending at the same place were dropped? If so, where could you start and end the drive?
5   In chess a knight's move always takes him two squares along one row or column of squares and one square in a perpendicular direction, as shown in Fig. 18.10, where the knight's possible moves are to the squares marked with X.
   (*a*) Is it possible to move a knight around an empty chessboard in such a way that it ends where it began and along the way makes just one move between any two squares where such a move

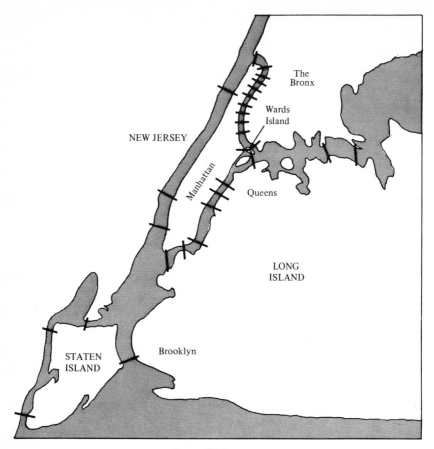

Figure 18.9

is possible? This amounts to asking for an Eulerian path in a network where 64 vertices correspond to the 64 squares on the chessboard. *Hint*: How many moves can the knight make from square A?

(*b*) Over a century ago chess buffs showed that it is possible to move the knight around the board so that it lands on every square exactly once and ends up where it began. What kind of path does this amount to?

*6 Maps are usually printed in several colors so that adjoining countries have different colors. For that the map in Fig. 18.11 needs four colors, since each country borders the other three. Does any map on a plane need five colors? Experiment a bit with this question.

   This problem was raised over a century ago, but, despite a great deal of effort, nobody knows to this day if such a map could exist! For purposes of the problem, regions which meet at isolated points, like Arizona and Colorado, are not considered as having a common border. We also rule out divided countries, like the continental United States with Alaska, Pakistan before Bangladesh, or West Germany with West Berlin. Water, which acts like a divided country, is also ignored. Strange to say, this problem has been completely solved for maps on other surfaces, such as those shaped like a doughnut. At heart it turns out to be a network problem. Can you see why?

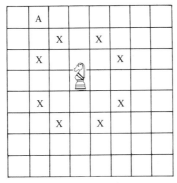

Figure 18.10

Figure 18.11

18.4 ONE-SIDED SURFACES

Figure 18.12 shows a *Möbius band*, which is curious in that it has only one side and one edge.[3] You can see that for yourself by running your finger along such a band, which is easily made by giving a half twist to a strip of paper and then taping the ends together. What happens if you cut a Möbius band down the center (along the dashed line in Fig. 18.12)? Try to predict the results, and then make a Möbius band and cut it. As a second experiment, what will result from cutting a Möbius band not down its center but along a line one-third of the way across? Try it and see!

An even more surprising result occurs if we first make a one-sided surface by putting three half-twists into a strip of paper and then fastening the ends, as in Fig. 18.13, and then cut it down the center.

[3]Ferdinand Möbius (1790–1868) was a German mathematician.

Figure 18.12

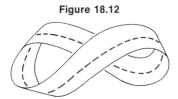

Figure 18.13

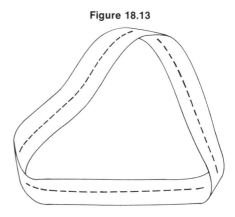

The result is a single two-sided surface twice as long as the original one-sided surface—with a knot in it!

What are the properties of one-sided surfaces made from taping strips of paper with five, seven, and nine half twists into bands? What patterns are there? What further questions do they raise?

We can not pursue these matters here. We have introduced you to some important mathematical ideas and to the creative process of raising and trying to answer questions, which is the heart of mathematics. We hope we are leaving you with more curiosity and more unanswered questions than you had when you began.

APPENDIX
SETS

The theory of sets, prominent in many "new math" books, has been omitted here, because it involves a relatively large dose of vocabulary and symbolism, which we sought to play down, and because we did not need set theory in our informal treatment. But the concept of set is a basic one in mathematics, and since some knowledge of set theory will help you understand other mathematics books, the main ideas of set theory, as first developed by Cantor[1] in the nineteenth century, are presented here. We begin with a simple but practical problem.

A.1 AN INVENTORY PROBLEM

In a record shop a detailed inventory of the stock is being made. They find that 783 of the 2,157 recordings on hand are tapes, while the rest are disks. The stock has also been classified and counted by subject matter, such as jazz, rock, and classical music, and the question is how many disk records in the shop are of subjects other than classical music. It is known that there are 238 classical records on hand, but this includes both tapes and disks.

One approach would be to count the nonclassical disks, but this would be tedious as there are so many. Luckily, the clerk knows some set theory. He begins by counting the classical tapes, since there are few of them. He finds there are 27 classical tapes. How does this help find the number of nonclassical disks? You may enjoy grappling with this before reading on, but you are not expected to be able to solve it at this stage.

---

[1]Georg Cantor (1845–1918) was born in St. Petersburg (now Leningrad) but lived in Germany after 1856. He taught at the University of Halle from 1869 to 1905 and did his most important work there. His study of sets with infinitely many members led him to some remarkable conclusions, many of which he reached by extremely elegant reasoning. At first Cantor's work was viciously criticized by prominent mathematicians (which no doubt contributed to the mental breakdown he later suffered), but by the turn of the century its importance was widely recognized.

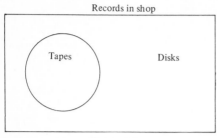

Figure A.1

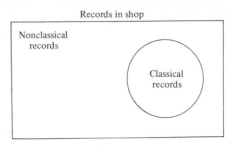

Figure A.2

We may picture the classifications of the records in the shop as in Figs. A.1 and A.2. Using the two classifications at the same time divides the records into four groups, as shown in Fig. A.3:

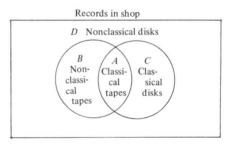

Figure A.3

Group *A*: classical tapes
Group *B*: nonclassical tapes
Group *C*: classical disks
Group *D*: nonclassical disks

Each record in the shop belongs to just one of these four groups, and our problem is to find how many belong to *D*. One way to do this is to find out first how many records belong to groups *A*, *B*, and *C*, since the rest are in *D*. There are 783 tapes, of which 27 are classical; so group *B* has 756 nonclassical tapes. There are 238 classical records, 27 of which are tapes; so group *C* consists of 211 classical disks. Then:

> Group *A* (classical tapes) has 27 records
> Group *B* (nonclassical tapes) has 756 records
> Group *C* (classical disks) has 211 records
> Total in sets *A*, *B*, *C*: 994 records

Since there are 2,157 records in all, the remaining $2,157 - 994 = 1,163$ must be nonclassical disks.

Exercises A.1

1   From the diagrams, what can you say about x in each case? For example in (c) we see that x is a reader of *Time* but not of *Fortune*, since x lies inside one circle and outside the other.

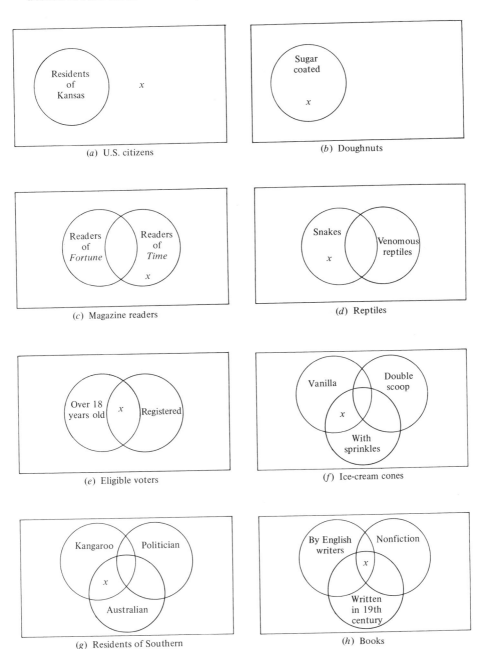

(a) U.S. citizens

(b) Doughnuts

(c) Magazine readers

(d) Reptiles

(e) Eligible voters

(f) Ice-cream cones

(g) Residents of Southern Hemisphere

(h) Books

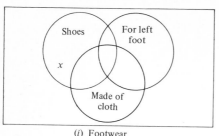

(i) Footwear

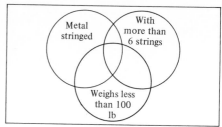

(j) Stringed musical instruments

2 Here are some objects and diagrams. Put each object into the area of every diagram which best fits it:

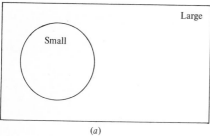

(a)

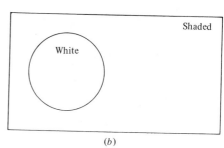

(b)

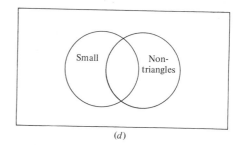

(c)

(d)

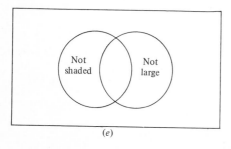

(e)

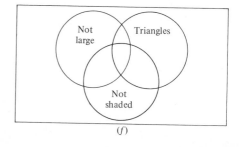

(f)

3 Put the numbers 1, 2, 3, 4, 5, 6, 7, 8, 9, 10 into each of these diagrams in the most appropriate way.

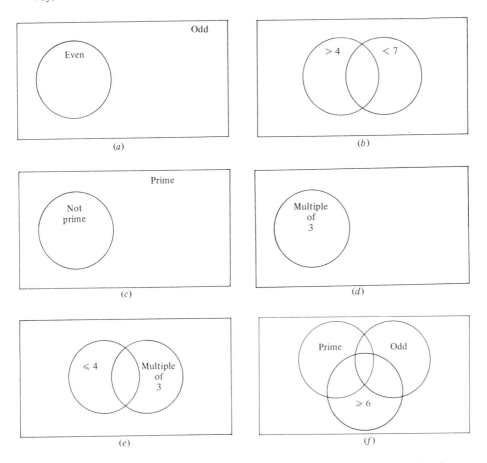

4 Twenty boys try out for baseball. Some have gloves, some have bats, some have both, and some have neither. What can you say about those who belong in these areas in Fig. A.4?
(*a*) The area with horizontal but not vertical lines.
(*b*) The area shaded both horizontally and vertically.
(*c*) The area shaded vertically but not horizontally.
(*d*) The unshaded area.

Figure A.4

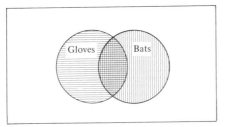

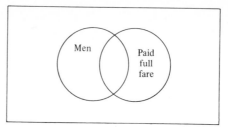

Figure A.5

5 If in Prob. 4 we learn that 14 of the 20 boys have gloves, 8 have bats, and 5 have both, how many have:
   (*a*) Bats but not gloves?
   (*b*) Gloves but not bats?
   (*c*) Neither bats nor gloves?

6 On a certain airplane flight the passengers (all adults) are classified by sex and by whether or not they paid full fare for their tickets. Thus we have four passengers types:

   Men who paid full fare
   Women who paid full fare
   Men who did not pay full fare
   Women who did not pay full fare

   (*a*) Locate each of these types in the Fig. A.5.
   (*b*) If there are 57 men on board and 28 of the passengers, including 19 men, did not pay full fare, and if there are 104 passengers altogether, how many women did not have reduced-fare tickets?

7 There are 1,171 books in a store, of which 837 are paperbacks and the rest hardbacks; 294 of the books, including 173 paperbacks, are fiction. How many nonfiction hardbacks are in the store? Make a diagram to show your work.

8 Of 453 people at a convention 270 were women, and 310 of those who attended were over twenty-five years old. If there were 70 men over twenty-five years old at the convention, how many of the women were twenty-five years old or less?

9 When 300 sixth graders took tests in reading and mathematics, 180 of them scored at or above grade level in reading and 137 of them scored at or above grade level in mathematics. Just 62 scored at or above grade level in both subjects. How many scored below grade level in both subjects?

A.2 SETS AND HOW THEY ARE DESCRIBED

The inventory problem happened to be about records, but the ideas used to solve it apply to any classification problem whatever. Any objects we choose to group or classify together form what is called a *set*, and the individual objects of which a set is formed are called its *elements* or *members*.

Since it does little good to think about sets if the thoughts cannot be communicated, a standard notation for describing sets is needed. One way is to list the members of the set in braces. (Children who find braces hard to make should use something easier like square brackets.) In this way of describing a set, the order in which the elements are listed is irrelevant, as are repetitions. For example,

{Maine, Texas, Iowa}

{Texas, Maine, Iowa}

and

{Maine, Texas, Maine, Iowa, Iowa}

are three ways of describing the same set.

Some sets, like those in the inventory problem, have too many members to list conveniently. They are best described by means of a property which characterizes their members, that is, distinguishes members of the set from nonmembers. In the standard notation for this, the set of records discussed in the inventory problem is $\{x : x$ is a record in the shop$\}$, read "the set of all x such that x is a record in the shop." The idea of this notation is to put, after the colon, the condition which the symbol x must satisfy in order to be a member of the set. The elements of the set are just those objects x for which the statement after the colon is true. For example, the statement "x is a whole number between 3.5 and 6.5" is true if x is 4, 5, or 6 but false if x is anything else; so $\{x : x$ is a whole number between 3.5 and 6.5$\}$ means $\{4,5,6\}$. Of course, any letter could be used in place of x here.

In describing a set this way care must be taken to state the criterion for membership clearly. For example, $\{z : z$ is a large number$\}$ does not describe a set, because the word "large" is relative and one cannot be certain just which numbers are supposed to be in the set. Even if the criterion for membership in a set is clear, there remains the question of just what objects actually satisfy the requirement. For example, $\{w : w$ is a prime number$\}$ is a clearly specified set, but no practical way of identifying its larger members is known. It is even possible to describe sets that have no members, such as

$\{a : a$ is an airplane that seats 5,000 people$\}$

or

$\{c : c$ is a city with 100 million inhabitants$\}$

Sets like these with no members are said to be *empty*. At first the idea of any empty set may seem strange; you may question the value of the concept, just as a youngster questions the number 0. How can there be a number of people in an empty room? The number is 0. How can we speak of the set of people in an empty room? The set is empty. In time the value of this concept, like that of 0, becomes clear.

Exercises A.2

1   List the elements of these sets.
   (a) $\{x : x$ is an integer between 0.4 and 5.7$\}$
   (b) $\{y : y + 1 = 11\}$
   (c) $\{z : 3z + 7 = 5z - 4\}$
   (d) $\{p : p$ is a prime less than 15$\}$
   (e) $\{x : x$ has been president of the United States some time since 1950$\}$

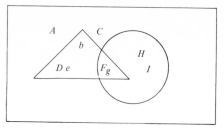

Figure A.6

2 Which of these are well defined sets? Justify your answers:
  (a) {x : x is a great musician}
  (b) {y : y is a female registered to vote in Alaska}
  (c) {n : n is an overpopulated nation}
  (d) {d : d is a dictatorship}
  (e) {x : x played for the 1927 New York Yankees}
3 Referring to Fig. A.6, list the elements of these sets:
  (a) {x : x is a capital letter in the circle}
  (b) {x : x is a consonant outside the triangle}
  (c) {x : x is in the triangle or outside the circle}
  (d) {x : x is a small letter or vowel inside the circle}
  *(e) {x : x is in neither the circle nor the triangle or x is in both the circle and the triangle}

A.3 SUBSETS

In our inventory problem we divided the records in the shop into smaller sets, such as the set of tapes in the shop, the set of nonclassical records in the shop, etc. These smaller sets are called *subsets* of the larger set which contains them. Figure A.7 shows set A as a subset of B, because as it is drawn no member of A could possibly lie outside of B. More formally, we say that set A is a subset of set B if each member of A is also a member of B. This amounts to saying that no members of A are outside B.

As a second example, consider the relations of {g,h} to {g,h,k,l} and to {h,i,j}. Figure A.8 shows both g and h inside the circle containing {g,h,k,l}, so

Figure A.7

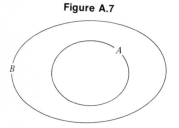

Figure A.8

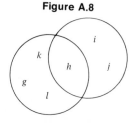

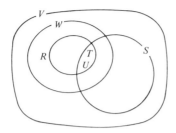

Figure A.9

$\{g,h\}$ is a subset of $\{g\ h,k,l\}$, but g lies outside the circle containing $\{h,i,j\}$, so $\{g,h\}$ is not a subset of $\{h,i,j\}$.

We write $A \subseteq B$ or $B \supseteq A$ to say that A is a subset of B. If $A \subseteq B$, must B contain any elements which are not in A? Not necessarily. $A \subseteq B$ means that each member of A belongs to B, but it is still possible that at the same time each member of B belongs to A, so that $B \subseteq A$. In that case A and B have the same members, and we write $A = B$, meaning A is the same set as B. For sets, as for numbers, "equals" means "is the same as."

A special case of these ideas arises in connection with an empty set (one with no members). An empty set is a subset of any set whatever, since there is no member of an empty set which lies outside any other set. In particular, this means that all empty sets are subsets of each other, so they are all equal. (This is reasonable enough; after all, a set of no horses has the same members as a set of no cows.) Therefore it is customary to speak of *the* empty set and to denote it by a special character $\emptyset$. which looks a bit like 0.

Exercises A.3

1 (a) In Fig. A.9 of which sets is W a subset?
   (b) Of which set is R a subset?
   (c) Of which sets is S a subset?
   (d) Of which sets is T a subset?
2 List the subsets of $\{p,q\}$ with:
   (a) No members   (b) One member   (c) Two members
3 List the subsets of $\{p,q,r\}$ with:
   (a) No members   (b) One member   (c) Two members   (d) Three members
4 List the subsets of $\{p,q,r,s\}$ with:
   (a) No members   (b) One member   (c) Two members
   (d) Three members   (e) Four members
5 From Probs. 2 to 4 we see that:
   (a) A set with two members has ____ subsets.
   (b) A set with three members has ____ subsets.
   (c) A set with four members has ____ subsets.
   (d) Guess the rule in parts (a), (b), and (c). How many subsets has a set with 10 members?
6 (a) Use your answers to Probs. 2 to 4 to complete this table:

Set	No. of subsets with 0 elements	No. of subsets with 1 element	No. of subsets with 2 elements	No. of subsets with 3 elements	No. of subsets with 4 elements
$\{p\}$					
$\{p,q\}$					
$\{p,q,r\}$					
$\{p,q,r,s\}$					

(*b*) Do you recognize the pattern in this table? *Hint*: See part (*a*) of Prob. 12 in Exercises 5.1.

A.4 COMBINING SETS

Instead of dealing with sets as isolated entities, we usually consider them in relation to each other, often forming new sets from the elements of sets already on hand. Here we consider the two main ways to do this.

Perhaps the simplest way to combine sets to form a new one is to pool their members, thus merging the various sets into a single set. This new set is called the *union* of the sets which were merged, since it is the result of uniting them. The symbol for the union of sets is $\cup$, which looks like the letter U and is used this way:

$$\{a,b,c,d\} \cup \{b,c,f\} = \{a,b,c,d,f\}$$

The shaded area of Fig. A.10 represents $H \cup K$. Note that the presence of a third set J has nothing to do with the question, which is simply to shade those areas which are in either the circle for K or the circle for H or both.

The criterion for membership in a union of sets is very inclusive; it can be met by belonging to one or more of the sets being united. A more exclusive combination of sets is their *intersection*, which consists of the elements common to all the sets. This use of the word "intersection" is related to its use when we speak of the intersection of two streets. To be in the intersection of two streets is to be in both at once; to be a member of the intersection of several

Figure A.10

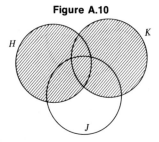

Figure A.11

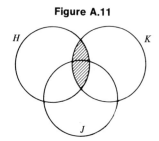

sets is to be a member of each and every one of them. The symbol for the intersection of sets and is ∩, which is merely a union symbol upside down.

$$\{a,b,c,d\} \cap \{b,c,f\} = \{b,c\}$$

and, analgous to Fig. A.10, Fig. A.11 shows $H \cap K$.

Exercises A.4

1 Describe the shaded areas in terms of unions and intersections.

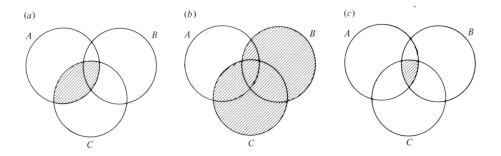

2 (*a*) Shade $A \cap B$:   (*b*) Shade $A \cap C$:

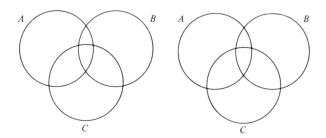

(*c*) Use the results of parts (*a*) and (*b*) to shade $(A \cap B) \cup (A \cap C)$:

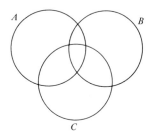

(*d*) Shade $B \cup C$:   (*e*) Shade $A \cap (B \cup C)$:

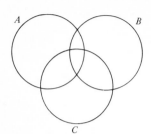   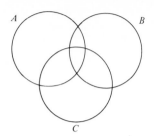

(*f*) Compare your answers from parts (*c*) and (*e*).

3   $P = \{\bigcirc, \triangle, \square, \bullet, \blacktriangle, \blacksquare\}$ $Q = \{\bullet, \blacktriangle, \blacksquare, \circ, \scriptstyle\triangle, \scriptstyle\blacksquare\}$
    $R = \{\bigcirc, \blacktriangle, \square, \bullet, \triangle, \square\}$

(*a*) Put each of the elements of these sets in the most appropriate place in Fig. A.12.
(*b*) $P \cup Q =$
(*c*) $P \cup R =$
(*d*) $(P \cup Q) \cap (P \cup R) =$
(*e*) $Q \cap R =$
(*f*) $P \cup (Q \cap R) =$
(*g*) How do your answers in parts (*d*) and (*f*) compare?

4   *A* has 5 members and *B* has 3 members; then:
(*a*) $A \cup B$ has at least ____ members.
(*b*) $A \cup B$ has at most ____ members.
(*c*) $A \cap B$ has at least ____ members.
(*d*) $A \cap B$ has at most ____ members.
(*e*) Fill in this table with *A* and *B* as above.

If $A \cup B$ has	5 members	6 members	7 members	8 members
Then $A \cap B$ has				

5   If *C* has 37 elements and *D* has 79:
(*a*) $D \cup C$ has at most ____ members.
(*b*) $D \cup C$ has at least ____ members.
(*c*) $D \cap C$ has at most ____ members.
(*d*) $D \cap C$ has at least ____ members.
(*e*) If $D \cup C$ has 83 members, how many has $D \cap C$?

Figure A.12

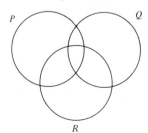

6 In the harbor there are 27 boats with sails and 19 with motors.
(*a*) There are at least ____ boats in the harbor.
(*b*) If there are 41 boats in the harbor and 4 have neither sails nor motors, how many have motors but not sails?

7 (You may want to refer to the sieve of Eratosthenes in Chap. 4.) Let
$P = \{x : x$ is a multiple of 2$\}$
$Q = \{x : x$ is a multiple of 3$\}$
$R = \{x : x$ is a multiple of 4$\}$
$S = \{x : x$ is a multiple of 5$\}$
Find simple descriptions for:
(*a*) $P \cap Q$ (*b*) $P \cap R$ (*c*) $Q \cap R$ (*d*) $P \cap S$ (*e*) $Q \cap S$
(*f*) $R \cap S$ (*g*) Can you generalize the pattern?

8 Let
$T = \{x : x$ is a factor of 168$\}$
$U = \{x : x$ is a factor of 72$\}$
$V = \{x : x$ is a factor of 105$\}$
List the members of these sets.
(*a*) $T \cap V$ (*b*) $T \cap U$ (*c*) $U \cap V$ (*d*) $T \cap U \cap V$
(*e*) In each case, your answers are all factors of a certain number. What characterizes it in every case?

*A.5 RUSSELL'S PARADOX

The power of set theory is in its generality, for in placing no restriction whatever on the nature of the objects of which sets are made up, Cantor developed a theory which could be applied in countless ways. This led to the hope that the single basic concept of set could be used as the foundation for all mathematical thought, simplifying and unifying it at the same time. Unfortunately, however, the very concept of set, which seemed so simple, was shown to be self-contradictory, at least in completely unrestricted form. This discovery, made in 1902, is known as *Russell's paradox.*[2]

Russell's paradox is based on the observation that the members of a set may themselves be sets. For example, if A, B, and C are sets, the set $\{A,B,C\}$ is a set whose members are sets. It is even possible to imagine sets which are members of themselves. One such is $D = \{x : x$ is a set with more than two members$\}$. Since there are more than two sets which have more than two members, D has more than two members, and so D is a member of itself.

Russell's paradox concerns the set E consisting of all sets which are not

---

[2]Bertrand Russell (1872–1970), one of the great thinkers of his time, wrote his most important mathematical work, "Principia Mathematica," with Alfred North Whitehead in 1910–1913, but his interests went far beyond mathematics, as the titles of a few of his dozens of books show. Among them are "Marriage and Morals" (he was married four times), "The Practice and Theory of Bolshevism" (he was a lifelong socialist and met Lenin and Trotsky, both of whom he called "evil," while traveling in Russia), "Education and the Social Order" (he founded a progressive school), "The ABC of Relativity," and "A Critical Exposition of the Philosophy of Leibniz." Russell was an outspoken pacifist, for which he was imprisoned in 1918. In the 1960s he led street demonstrations in England against America's Indochina policy, even though he was well into his nineties. This amazing man traveled extensively, lecturing for a year (1920–1921) in Peking and making many trips to the United States, including extended stays at Harvard, University of Chicago, and UCLA. He won the Nobel prize for literature in 1950.

members of themselves. Is E a member of itself? On the one hand E cannot be a member of itself, for if it were, then one of its members, namely itself, would belong to itself, and E is supposed to consist entirely of sets which do not belong to themselves. On the other hand, if E is not a member of E, then E is not a member of the set of sets which do not belong to themselves, so E is not a set which is not a member of itself. Therefore E must be a member of itself!

Several popularizations of Russell's paradox pose the dilemma in less theoretical terms. Russell himself made up one in 1919. Imagine a barber in a town who shaves all those (and only those) in town who do not shave themselves. Does the barber shave himself?

Of course there never was or could be such a barber. But Russell's paradox is not just a play on words or a result of faulty reasoning. On the contrary, it is inherent in a completely unrestricted concept of set. The paradox cannot be resolved, but it can be avoided; it does not arise unless sets refer to themselves, as E does, and no practical use of set theory involves such sets. The effort to clarify the concept of set has led to development of a technical subject known as *axiomatic set theory*. It justifies our naïve approach, provided we stay away from sets like E.

BIBLIOGRAPHY

If this book has succeeded, your contact with mathematics will not end here. You may find yourself teaching the subject, or you may just want to follow up ideas introduced here. The bibliography is designed to help you with this. It is not intended to be comprehensive (that would make it so big as to be nearly useless) but is a selected list of the author's personal favorites. In one way or another, each of these works shares the spirit of this book.

Aaboe, Asger: "Episodes from the Early History of Mathematics," Random House, New York, 1964. Chapter 3 is about Archimedes and his work.

Bell, Eric T.: "Men of Mathematics," Simon and Schuster, New York, 1937. Biographies of famous mathematicians including Gauss, Pascal, Descartes, Fermat, Newton, Euler, and Cantor.

Beiler, Albert H.: "Recreations in the Theory of Numbers," 2d ed., Dover, New York, 1966. A wealth of information about numbers in an informal, readable format. Trivial but fascinating curiosities are blended with informal discussions of deep questions and rich historical references. The discussion of perfect numbers and related topics is especially interesting, and chap. 8 is an excellent treatment of casting out 9s. Chapter 10 is on decimal fractions, and chap. 9 is on nondecimal numeration. Chapter 24 is on Fermat's last theorem.

Blumenthal, Leonard M.: "A Modern View of Geometry," Freeman, San Francisco, 1961. Chapter 1 discusses the nineteenth century rise of non-Euclidean geometry. Chapter 2 has a thorough discussion of sets in general and Russell's paradox in particular.

Choquet, Gustave: "Geometry in a Modern Setting," Hermann, Paris, 1969. Section 57, "Difficulties in Defining an Angle" is the basis for Sec. 16.1 of this book. Section 57 is only about a page long, is easy reading, and is well worth the effort of finding it.

Chrystal, George: "Textbook of Algebra," vol. 1, Dover, New York, 1961. A paperback reprint of a nineteenth century classic. Written in a mildly archaic style, this book is awesomely comprehensive. Much of it is advanced, but you will have no trouble with chap. 1, which has a careful discussion of commutativity, associativity, and distributivity. Probably the best existing reference book on algebra.

Copeland, Richard: "Diagnostic and Learning Activities in Mathematics for Children," Macmillan, New York, 1974. A very practical booklet for anyone seeking to apply Piaget's ideas in elementary schools.

————"How Children Learn Mathematics: Teaching Implications of Piaget's Research," 2d ed., Macmillan, New York, 1974.

Cundy, H. Martyn, and A. P. Rollett: "Mathematical Models," Oxford, New York, 1961. An excellent source of detailed instructions for building polyhedra and other models.

De Temple, Duane, and Jack Robertson: The Equivalence of Euler's and Pick's Theorems, *The Mathematics Teacher*, vol. 67, no. 3 (January 1974). A thorough discussion of the close relationship between Pick's theorem (Prob. 4 of Exercises 13.2) and Euler's formula, which is discussed in Sec. 18.2.

Escher, Maurits C.: "The Graphic Work of M. C. Escher," Hawthorn, New York, 1961. This collection of prints is not a mathematics book at all, but Escher's ingenious use of geometric concepts in art is unique and intriguing.

Eves, Howard: "An Introduction to the History of Mathematics," 3d ed., Holt, New York, 1969. A readable history of mathematics through the nineteenth century. Of special interest are chap. 1, on numeration; chap. 3, on Pythagoras and the Pythagoreans, chap. 5 on Euclid's "Elements," and sec. 4.6, "A Chronology of π."

Gardner, Martin: Mathematical Games, *Scientific American*. This monthly feature has long been a favorite with students, teachers, and recreational mathematicians. Here are some you might enjoy best, listed by topics (some have been published separately in collections):

> Binary system, May, 1957, December 1960, August 1962, February 1968
> Calculating prodigies, April 1967
> Calculating in base -2, April 1973
> Curves of constant width, February 1963
> Diagrams that make complex formulas clear at a glance, October 1973
> Digital roots, July 1958
> Divisibility rules, September 1962
> Fermat's last theorem, July 1970
> Fibonacci numbers, August 1959, December 1966, March 1969
> Four-color map problem, September 1960
> Hamiltonian circuits, May 1957
> Möbius band, June 1967, July 1963, December 1968
> Perfect numbers, March 1968
> Polyhedra, December 1958, September 1971
> Prime numbers, March 1964, March 1968, August 1970
> Pythagorean theorem, June 1960, October 1964, November 1971
> Ternary system (base 3), May 1964
> Tessellations, April 1961
> Unsolved problems in elementary number theory, December 1973

Gridgeman, N. T.: John Napier and the History of Logarithms, *Scripta Mathematica*, Vol. 29, nos. 1 and 2 (Spring-Summer, 1973). Of the many excellent sources on Napier (Eves' book listed above, for example) this one stands out for its thoroughness and clarity.

Holden, Alan: "Shapes, Space, and Symmetry," Columbia University Press, New York, 1971. A beautiful collection of photographs of polyhedron models.

Hogatt, Verner: "Fibonacci and Lucas Numbers," Houghton Mifflin, Boston, 1969. A leading member of the Fibonacci Association has assembled here some of the countless fascinating properties of Fibonacci numbers and their relatives.

Jacobs, Harold: "Mathematics: A Human Endeavor," Freeman, San Francisco, 1970. Aptly subtitled, "A Text for Those Who Think They Don't Like the Subject," this book fascinates all who pick it up. The material is based on Martin Gardner's columns in *Scientific American*, but it is simplified, profusely illustrated, and adorned with cartoons. Especially good on graphing, logarithms, scientific notation, polyhedra, topology, and introductory probability and statistics.

Jencks, Stanley M., and Donald M. Peck: "Building Mental Imagery in Mathematics," Holt, New York, 1968. An informal treatment of selected topics. Especially good on fractions, areas, and functions or rules.

"Mathematics in the Modern World, Readings from *Scientific American*," Freeman, San Francisco, 1968. This collection includes short biographies of Descartes, Newton, Hamilton, and Rumanujan (an especially interesting one). It has an expository article on paradox, which discusses Russell's and other paradoxes, and it also includes a translation of Euler's original paper, The Königsberg Bridges, which is unusually clear and readable.

Ogilvy, C. Stanley: "Tomorrow's Math," 2d ed., Oxford University Press, New York, 1972. A book of unsolved problems in elementary mathematics, written in a pleasant, informal style. Chapter 5 contains a rich collection of unsolved problems about prime numbers.

Olds, C. D.: "Continued Fractions," Random House, New York, 1963. Treats the topic more deeply that we did here (Sec. 15.4) but requires a knowledge of high school algebra.

Ore, Oystein: "Graphs and Their Uses," Random House, New York, 1963. Graph here means what we have called networks (see Chap. 18).

Polya, George: "Mathematical Discovery," Wiley, New York, 1965. Subtitled "On Understanding, Learning, and Teaching Problem Solving," this book is a storehouse of mathematical and pedagogical ideas. Volume I is filled with specific examples and many problems, some quite hard but magnificently chosen. Volume II is a more general discussion of learning and problem solving as processes, Chapters 14, On Learning, Teaching, and Learning Teaching, and 15, Guessing and Scientific Method, are especially interesting; they are not hard reading.

Ranucci, Ernest: Fruitful Mathematics, *The Mathematics Teacher*, vol. 67, no. 1 (January 1974). Based on Ranucci's observation of how street vendors in Ecuador pile fruit, this article relates figurate numbers to crystal-packing questions.

———Master of Tessellations: M.C. Escher, 1898–1972, *The Mathematics Teacher*, vol 67, no. 4 (April 1974). Describes Escher's work and how it grew out of studying the symmetry of tile patterns in the Alhambra.

Sawyer, W. W. : "Vision in Elementary Mathematics," Penguin, Baltimore, 1964. Sawyer's easy-going, chatty style makes this light reading, but it contains both a sound philosophy of education and concrete ideas on how to use it.

Spaulding, Raymond E.: Pythagorean Puzzles, *The Mathematics Teacher*, vol. 67, no. 2 (February 1974). An activity-oriented presentation of some puzzles akin to Prob. 12 of Exercises 14.2 in this book.

Spitznagel, Edward L.: "Selected Topics in Mathematics," Holt, New York, 1971. Chapter 5 is a careful discussion of the relation of map-coloring problems to networks.

Steinhaus, Hugo: "Mathematical Snapshots," 3d American ed., Oxford University Press, New York, 1969. By ingenious use of pictures, this book presents a wide variety of mathematical ideas clearly, yet has little printed material. This is not a text but should be kept on hand for browsing. Chapters 4, 7, 8, and 12 have interesting ideas about tessellations, polyhedra, crystal packing, and topology.

Stewart, Bonnie M.: "Adventures among the Toroids," published by the author, 4494 Wausau Road, Okemos, Mich. 48864, 1970. This delightful book, beautifully hand-lettered and illustrated by the author, grew out of his work with elementary school teachers, but it gets into some subtle and entirely new areas. Much of the material can easily be learned by anyone willing to make some of the beautiful toroid models described.

Teeters, Joseph L.: How to Draw Tessellations of the Escher Type, *The Mathematics Teacher*, vol. 67, no. 4 (April 1974).

Toth, Imre: Non-Euclidean Geometry before Euclid, *Scientific American*, vol. 221, no. 5 (November 1969). Remarkable evidence that Aristotle and other ancient Greeks developed ideas of non-Euclidean geometry.

Wentworth, Daniel F., J. K. Couchman, John C. MacBean, and Adam Stecher: "Mapping Small Places," Winston, Minneapolis, 1972. Beautifully written and illustrated, this small book teaches a great deal of geometry and relates it strongly to the real world. It is aimed at grades 4 to 6, but much of the material is easily adapted to other ages.

Wenninger, Magnus: "Polyhedron Models," Cambridge University Press, New York, 1971. The main feature of this book is photographs of polyhedron models, including some exceptionally beautiful but complex ones which are not often shown. Directions for building models are included, but they are not as full as those in Cundy and Rollett's "Mathematical Models," cited above.

Wirtz, Robert, Morton Botel, Max Beberman, and W. W. Sawyer: "Math Workshop for Children," rev. ed., Encyclopaedia Britannica Press, Chicago, 1967. This series consists of six workbooks (levels A to F correspond roughly to grades 1 to 6, but the series is actually nongraded), as well as supplementary materials. It is a goldmine of exciting ways of presenting mathematical ideas to elementary school children. Box puzzles are used extensively and imaginatively in the four lower levels. Highlights of level F include how to use angles in navigation, an empirical, activity-oriented treatment of π, and a beautiful section on Pick's theorem (Prob. 4 of Exercises 13.2 in this book).

ANSWERS TO SELECTED EXERCISES

Exercises 1.3

2 (a) 20,100   (c) 10,300   (e) 2,313
3 6,679,500 yards
4 1,000

Exercises 1.4

2 (a)

	4	7	9	20
8	32	56	72	160
3	12	21	27	60
2	8	14	18	40
13	52	91	117	260

Exercises 1.5

2 (a)

	30	8	38
50	1,500	400	1,900
4	120	32	152
54	1,620	432	2,052

(b)

	200	60	5	265
3	600	180	15	795
70	14,000	4,200	350	18,550
73	14,600	4,380	365	19,345

3 (a)

	600	50	4	654
100	60,000	5,000	400	65,400
3	1,800	150	12	1,962
103	61,800	5,150	412	67,362

Exercises 2.1

1 (*a*) We had a loss.
  (*b*) The meeting started 5 minutes ago.
  (*c*) The plane is dropping 200 feet a minute.
  (*d*) In a recession the economy may shrink.
  (*e*) The rocket is leaving the earth at 3 miles per second.
  (*f*) The north poles of two magnets repel each other.
  (*g*) A subsidy for poor people.
  (*h*) The quarterback lost 7 yards.
2 (*a*) ⁻3   (*b*) 3   (*c*) 3
  (*d*) The result is negative or positive according as the number of minus signs is odd or even.

Exercises 2.3

1 (*a*) 0   (*b*) 0   (*c*) 0
  (*d*) Any number and its opposite add up to 0.
2 (*a*) ⁻39   (*c*) ⁻47   (*e*) 1   (*g*) ⁻931   (*i*) ⁻58   (*k*) ⁻101.
3 (*b*)

⁻21	47	26
73	⁻52	21
52	⁻5	47

6 (*a*) 0   (*d*) 825

Exercises 2.5

1 (*a*) positive   (*c*) negative   (*e*) negative
2 (*a*)

	⁻11	5	⁻6
⁻8	88	⁻40	48
⁻2	22	⁻10	12
⁻10	110	⁻50	60

3 (*a*) ⁻42   (*c*) 42   (*e*) ⁻6   (*g*) 1

Exercises 3.1

1 36, 45, 55, 66, 78, 91, 105, 120
2 (*a*) 100   121   144   169   196   225   256   289   324   361   400

3 3   6   10   15   21   28   36   45   55   66   78   91   105   120   136   153

4 (*a*) 1   (*b*) 25   (*c*) 121
         1 is square of first odd number
        25 is square of third odd number
       121 is square of sixth odd number
                        ↑

              These are triangle numbers

5 (*b*) triangular
6 (*b*) 3   (*d*) 10

Exercises 3.2

1 The sums are tetrahedral numbers.
2 yes

0   1   8   27   64   125   216   343   512   729   1000

Last digits add to 10 or 0

3 These are cubes: $31 + 33 + 35 + 37 + 39 + 41 = 6^3$.
4 These are squares of the triangular numbers.
5 These are tetrahedral numbers.
6 (*a*) 27   (*c*) 125
　　(*e*) We predict $35 + 4 \cdot 56 + 84 = 7^3$ and $56 + 4 \cdot 84 + 20 = 8^3$.
7 (*a*) 36 total   (*b*) 100
　　(*c*) If the square has *n* boxes on a side, the total number of rectangles is $(1 + 2 + 3 + \cdots + n)^2$.
9 (*a*) squares   (*b*) tetrahedral numbers
11 (*b*) tetrahedral numbers, . . . tetrahedral numbers.

Exercises 4.2

1

```
        360
       /  \
    [2]· 180
     /  \
    2 · 15 · [12]
   / \  / \  / \
  2 · 3 · 5 · 2 ·[6]
 / \ / \ / \ / \ / \
2 · 3 · 5 · 2 ·[2]·[3]
```

3 (*a*) prime   (*b*) $5 \cdot 5$ (*c*) $3 \cdot 19$ (*d*) prime
　　(*e*) $3 \cdot 29$ (*f*) $7 \cdot 13$ (*g*) prime   (*h*) $2 \cdot 2 \cdot 2 \cdot 2 \cdot 2 \cdot 2 \cdot 2$
4 $1 \cdot 2 \cdot 3 \cdot 4 \cdot 5 \cdot 6 \cdot 7$

Exercises 4.3

2 (*a*) 8   (*c*) 5   (*d*) 5   (*e*) 3
3 13. The next higher prime, 17, has a square of 289, higher than the sieve goes.
5 4, 9, 25, 49, and 121 are the smallest such examples. All are squares of primes.

Exercises 5.1

1 (*a*) 3^5 (*c*) $(^-2)^6$ (*e*) $2^2 \cdot 3^3 \cdot 7^2 \cdot 11$
2 (*c*) In general x^y is not the same as y^x.
3 (*e*) Any power of 1 is 1.
4 (*a*) 5   (*c*) 927
5 (*a*) 10   (*c*) 1,000   (*d*) 10,000
6 (*b*) 8   (*c*) 32
7 2, 4, 8, and 16
8 (*b*) The leftmost number in each row is a square number.
　　(*c*) You get triangular numbers.
　　(*d*) $25 + 26 + 27 + 28 + 29 + 30 = 31 + 32 + 33 + 34 + 35$

9 (*b*) The numbers being squared are alternate triangular numbers.

(*c*) They are 4 times triangular numbers.

(*d*) $36^2 + 37^2 + 38^2 + 39^2 + 40^2 = 41^2 + 42^2 + 43^2 + 44^2$

10 $1^3 + 12^3 = 9^3 + 10^3$

11 (*b*) $1 + (2^3 + 3)^2$　　　(*c*) $[(1 + 2)^3 + 3]^2$

12 (*e*) You get Pascal's triangle.

(*f*) The next two rows are

1　　8　　28　　56　　70　　56　　28　　8　　1

1　　9　　36　　84　　126　　126　　84　　36　　9　　1

(*g*)　　　$1^2 + 1^2 = 2$

$1^2 + 2^2 + 1^2 = 6$

$1^2 + 3^2 + 3^2 + 1^2 = 20$

$1^2 + 4^2 + 6^2 + 4^2 + 1^2 = 70$

$1^2 + 5^2 + 10^2 + 10^2 + 5^2 + 1^2 = 252$

In general the sum of the squares of the numbers in one row is the middle number in the row twice as far down the triangle.

13 (*a*) 2, 5, 13, 34, 89, 233. These are alternate Fibonacci numbers.

(*d*) The sums are the Fibonacci numbers.

14 2,801

15 (*b*) four

Exercises 5.2

1 (*a*) $2 \cdot 2 \cdot (2 \cdot 2 \cdot 3) \cdot 3 \cdot 5$　　　(*c*) $(2 \cdot 2 \cdot 2 \cdot 2) \cdot 3 \cdot 5$

3 (*a*) $3^2 \cdot 5^2 = 225$　　　(*c*) $2^2 \cdot 3 = 12$

4 (*a*) 112　　　(*c*) 60　　　(*e*) 216　　　(*g*) 175　　　(*i*) 840　　　(*k*) 61,200

5 (*a*) $2^4 \cdot 7^3$　　　(*c*) $2^3 \cdot 3^4 \cdot 5^7 \cdot 11^4$

Exercises 5.3

2 (*a*) 28　　　(*b*) 496　　　(*c*) 8,128

(*d*) $1^3 + 3^3 + 5^3 + \cdots + 31^3$ should be the next perfect number according to the pattern, but it is not. All known perfect numbers are sums of odd cubes of this form, but not all such sums yield perfect numbers.

4 (*b*) 284

Exercises 6.1

2 (*b*)

8	5	13
3	11	14
11	16	27

(*d*)

35	22	57
⁻7	⁻9	⁻16
28	13	41

4 (*a*) 14,777 feet　　　(*b*) $14.495 - {}^-282$

6 (*b*) $^-27 - {}^-47$

Exercises 6.2

1 (*a*) $7 + {}^-1$　　　(*c*) $15 + {}^-15$　　　(*e*) $0 + {}^-18$

(*g*) $^-12 + {}^-13$　　　(*i*) $123 + 38$　　　(*k*) $11 + 8$

2 (*a*) 6　　　(*c*) 0　　　(*e*) ⁻18　　　(*g*) ⁻25　　　(*i*) 161　　　(*k*) 19

3 (*a*) no　　　(*b*) no

4 (b) 0   1   8   27   64   125   216
     1   7   19   37   61   91
       6   12   18   24   30
         6   6   6   6

(d) For squares the second differences were all the same, and for cubes the third differences were all the same. We predict in this case the fourth differences will all be the same. The constant second differences for squares were 2, and the constant third differences for cubes were 6. Since $2 = 1 \cdot 2$ and $6 = 1 \cdot 2 \cdot 3$, we predict constant fourth differences of $1 \cdot 2 \cdot 3 \cdot 4 = 24$ for fourth powers.

5 (a) 4 and 4   (c) ⁻17 and ⁻17

Exercises 6.3

1 (a) $93 = \underline{90} + 3 = \underline{80} + 13$
    $57 = 50 + 7 = \underline{50} + 7$
                $\underline{30} + \underline{6} = 36$

(c) $235 = 200 + \underline{30} + 5 = 100 + \underline{130} + 5$
   $182 = \underline{100} + 80 + 2 = 100 + \underline{80} + 2$
                         $\underline{50} + \underline{3} = \underline{53}$

2 (a) 3   (c) 853   (e) 8   (g) 218

Exercises 6.4

1 (a)

	4	7	11
5	20	35	55
2	8	14	22
7	28	49	77

(c)

	3	2	5
⁻2	⁻6	⁻4	⁻10
9	27	18	45
7	21	14	35

(e)

	1	4	0	5
3	3	12	0	15
13	13	52	0	65
16	16	64	0	80

2 Division is neither commutative nor associative.

3 (a) positive   (b) negative   (c) negative   (d) positive

Exercises 6.5

1 (c) If the square of any odd integer is divided by 4, the remainder is 1. If the square of any even integer is divided by 4, the remainder is 0.

2 (b) Those which leave remainder 1 can each be expressed as the sum of two squares. The others cannot be so expressed.

(e) To be the sum of two squares an odd prime must be the sum of an odd square and an even square, so it must be 1 more than a multiple of 4. It is also true, though harder to show, that *every* prime which is 1 more than a multiple of 4 can be expressed as a sum of two squares.

6 (d) You should have the 3-digit number you began with.

Exercises 7.1

1 (a) $\dfrac{11}{3}$ (c) $\dfrac{15}{-4}$

2 (*a*) $2 \div 3$ (*c*) $-14 \div 21$

3 (*a*) $\dfrac{1}{3}$ (*c*) $\dfrac{3}{12}$ (*e*) $\dfrac{5}{9}$ (*g*) $\dfrac{9}{9}$

6 (*a*) 1   (*c*) 1   (*d*) -1 (*e*) 5   (*f*) 3

Exercises 7.2

1. (*a*)

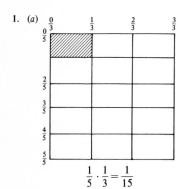

$$\frac{1}{5} \cdot \frac{1}{3} = \frac{1}{15}$$

(*b*)

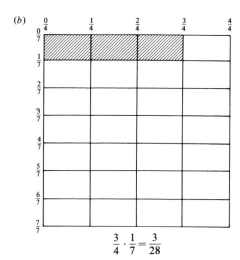

$$\frac{3}{4} \cdot \frac{1}{7} = \frac{3}{28}$$

(*c*)

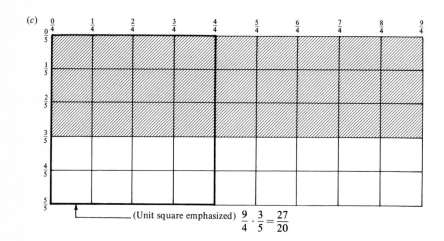

(Unit square emphasized) $\dfrac{9}{4} \cdot \dfrac{3}{5} = \dfrac{27}{20}$

(d)

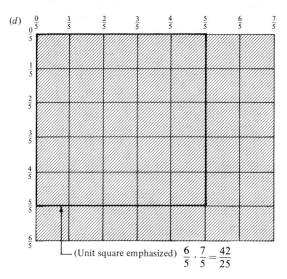

└─ (Unit square emphasized) $\dfrac{6}{5} \cdot \dfrac{7}{5} = \dfrac{42}{25}$

2 (a) $\dfrac{35}{90}$ (c) $\dfrac{4,700}{10,353}$ (d) $\dfrac{pr}{qs}$

Exercises 7.3

1 (a) 2   (b) −3   (c) 50
2 (a) 7   (b) 10   (c) −13
3 (a) $\dfrac{-5}{12}$ (c) $\dfrac{18}{-11}$
4 (a), (c), and (d)

Exercises 7.4

1 (a) $\dfrac{1}{3}$ (c) $\dfrac{1}{3}$

2 $\dfrac{1}{8}$

3 (a) $\dfrac{2}{5}$ (c) $\dfrac{4}{7}$

4 (a) $\dfrac{2}{5}$ (c) $\dfrac{4}{7}$

5 (a) $\dfrac{-2}{18}$ or $\dfrac{-1}{9}$ (c) $\dfrac{-1}{10}$

6 (a) 1   (c) 1

     (e) $\dfrac{a}{b} \cdot \dfrac{b}{a} = 1$, no matter what a and b are (provided neither is 0).

7 (a) $\dfrac{5}{3}$ (c) $\dfrac{7}{-9}$

8 (a)
$$
\begin{array}{r}
1\ R\ 18 \\
150\overline{)168} \\
\underline{150} \\
18
\end{array}
\qquad
\begin{array}{r}
8\ R\ 6 \\
18\overline{)150} \\
\underline{144} \\
6
\end{array}
\qquad
\begin{array}{r}
3\ R\ 0 \\
6\overline{)18} \\
\underline{18} \\
0
\end{array}
$$

The last nonzero remainder, 6, is the highest common factor of 150 and 168.

Exercises 7.5

1 (a) $\dfrac{3}{7} \div \dfrac{4}{5} = \dfrac{\frac{3}{7}}{\frac{4}{5}} = \dfrac{\frac{3}{7}}{\frac{4}{5}} \cdot \dfrac{\frac{5}{4}}{\frac{5}{4}} = \dfrac{\frac{15}{28}}{\frac{20}{20}} = \dfrac{\frac{15}{28}}{1} = \dfrac{15}{28}$

(c) $\dfrac{-3}{11} \div \dfrac{7}{15} = \dfrac{-\frac{3}{11}}{\frac{7}{15}} = \dfrac{-\frac{3}{11}}{\frac{7}{15}} \cdot \dfrac{\frac{15}{7}}{\frac{15}{7}} = \dfrac{-\frac{45}{77}}{\frac{105}{105}} = \dfrac{-45}{77}$

2 (a) $\dfrac{20}{8}$ (c) $\dfrac{3}{2}$ (e) $\dfrac{-15}{16}$

(g) If $\dfrac{r}{s} \div \dfrac{t}{u} = \dfrac{x}{y}$, then $\dfrac{t}{u} \div \dfrac{r}{s} = \dfrac{y}{x}$ provided t and r are not zero.

Exercises 7.6

1 (a) $\dfrac{3}{4}$ (c) $\dfrac{15}{16}$ (e) $\dfrac{1}{2} + \dfrac{1}{4} + \cdots + \dfrac{1}{2^n} = \dfrac{2^n - 1}{2^n}$

2 (a) $\dfrac{4}{9}$ (c) $\dfrac{40}{81}$

3 (a) $\dfrac{5}{6}$ (c) $\dfrac{9}{20}$ (e) In general, $\dfrac{1}{n} + \dfrac{1}{n+1} = \dfrac{2n+1}{n(n+1)}$

4 (a) $\dfrac{2}{3}$ (c) $\dfrac{2}{15}$ (e) $\dfrac{1}{(n-1)n} + \dfrac{1}{n(n+1)} = \dfrac{2}{n^2 - 1}$

5 (a) 0   (c) 0

6 (a) $\dfrac{5}{24}$ (b) $-\dfrac{1}{10}$ (c) $\dfrac{76}{77}$

7 (a) $1\frac{5}{12}$ (c) $16\frac{2}{3}$

8 (a) $\dfrac{5}{3}$ (b) $\dfrac{55}{8}$

9 $133\frac{1}{3}$

11 (a) $\dfrac{1}{2}$ (c) $\dfrac{1}{4}$ (e) $\dfrac{1}{6}$ (f) $\dfrac{1}{19}$ and $\dfrac{1}{49}$

12 (a) $\dfrac{1}{2}$ (c) $\dfrac{1}{4}$ (e) $\left(1 - \tfrac{1}{2}\right)\left(1 - \tfrac{1}{3}\right)\left(1 - \tfrac{1}{4}\right) \cdots \left(1 - \tfrac{1}{n}\right) = \dfrac{1}{n}$

13 (a) $\dfrac{3}{4}$ (c) $\dfrac{5}{8}$ (e) $\left(1 - \tfrac{1}{2^2}\right)\left(1 - \tfrac{1}{3^2}\right)\left(1 - \tfrac{1}{4^2}\right) \cdots \left(1 - \tfrac{1}{n^2}\right) = \dfrac{n+1}{2n}$

14 (a) $\dfrac{5}{3}$ (c) $\dfrac{5}{5}$

Exercises 8.1

1 (a) I   (b) III   (c) II   (d) IV   (e) On the y-axis

Exercises 8.2

1 (a) Add 2 to the number   (c) Square the number, then add 1.
(e) Add 6 to the number, then divide the result by 3 less than the number.

2 (a)

1	5	13	20	0	−2
7	11	19	26	6	4

Shorthand for rule C: $n + 6$

3 (*a*)

3	0	5	2	10	19	50	−3	−28
6	0	10	4	20	38	100	−6	−56
12	6	16	10	26	44	106	0	−50

(*b*) $2x + 6$

4 (*b*) $2(n + 6)$ or $2n + 12$

6 (*b*) n

7 (*a*) subtracts 10 (or adds −10)

(*c*) divides by 7 $\left(\text{or multiplies by } \dfrac{1}{7}\right)$

(*e*) adds 4

10 The number of feet fallen is about $16t^2$, where t is the time in seconds since the object was dropped.

11 (*b*) The rule is n lines from each of two corners cut the triangle into $(n + 1)^2$ pieces.

(*e*)

0	**1**	**2**	**3**
0	3	12	27
0	3	6	9
1	1	1	1

Exercises 8.3

1 (*a*)

Inputs n	6	5	3	1	0	−1	−2
Outputs $3n - 7$	11	8	2	−4	−7	−10	−13

(*b*) $(6, 11), (5, 8), (3, 2), (1, -4), (0, -7), (-1, -10), (-2, -13)$

(*c*) See p. 236 for answer

2 (*a*)

Inputs n	−3	−1	−$\frac{1}{4}$	0	$\frac{1}{2}$	1	2
Outputs $2n + \frac{1}{2}$	−$5\frac{1}{2}$	−$1\frac{1}{2}$	0	$\frac{1}{2}$	$1\frac{1}{2}$	$2\frac{1}{2}$	$4\frac{1}{2}$

(*b*)

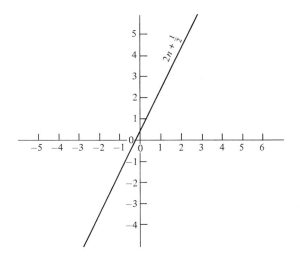

1 (*c*)

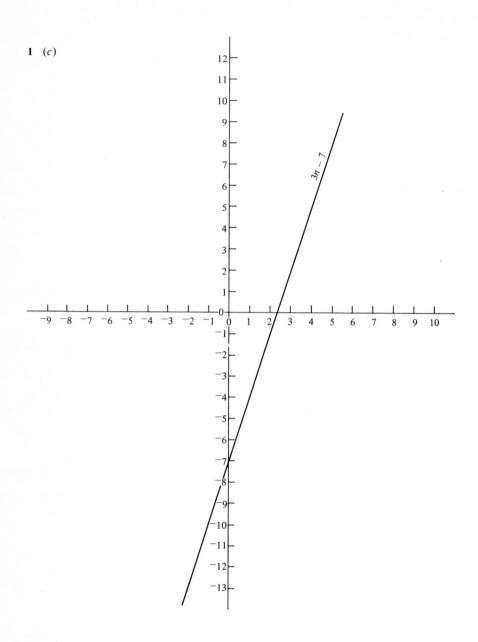

3 (*b*) $(-3,9)$, $(-2,4)$, $(-1,1)$, $(0,0)$, $(1,1)$, $(2,4)$, $(3,9)$

(*d*) $\left(-2\frac{1}{2}, 6\frac{1}{4}\right), \left(-1\frac{1}{2}, 2\frac{1}{4}\right), \left(-\frac{1}{2}, \frac{1}{4}\right), \left(\frac{1}{2}, \frac{1}{4}\right), \left(1\frac{1}{2}, 2\frac{1}{4}\right), \left(2\frac{1}{2}, 6\frac{1}{4}\right).$

4

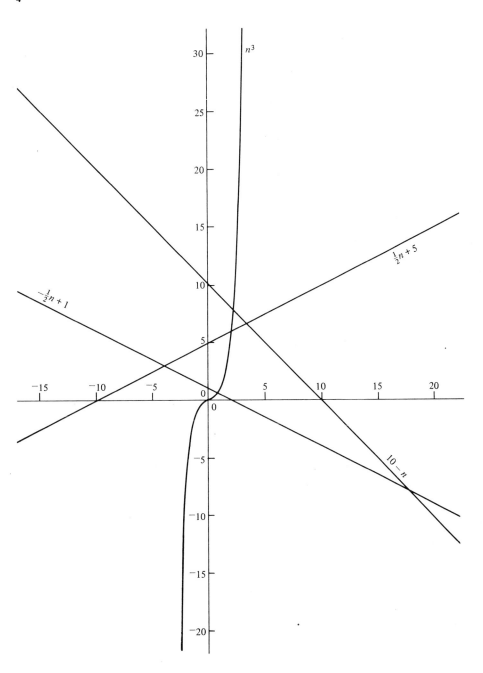

Exercises 9.1

1 (*a*) yes   (*b*) yes   (*c*) no   (*d*) yes   (*e*) yes   (*f*) no
2   4 and 5

Exercises 9.2

1 (*a*) ⅄⅄○○○○○   (*c*) ⅄⅄⅄⅄⅄○
2 (*a*) ⅄⅄⅄ ○○○○ = ⅄⅄⅄⅄⅄⅄⅄○○
3 (*a*) $3x + 2 = 5x$ (*b*) $7x + 5 = 2x + 15$
4 (*a*) ○○○   (*c*) ○○○○○   (*d*) $15 + 13x$
    (*f*) $24t + 36$ (*h*) $2r + 3$
5 (*a*) ○ = ⅄⅄⅄ ○○○
6 (*a*) $3s = 4s + 2$ (*c*) $m + 3 = 2m + 2$
7 (*a*) subtract 5   (*d*) divide by 5
8 $a = 1,\ c = 3,\ e = 3,\ g = \frac{3}{4}$
9 (*a*) $2y$ (*c*) $3k + 5$
10 (*a*) 16
11 (*a*) 1   (*c*) 4   (*e*) 4   (*g*) $\frac{7}{3}$

12 (*a*) $y = \frac{7}{5}$

13 (*a*) 14   (*c*) -6 (*e*) -4 (*f*) $\frac{11}{5}$

14 (*b*) $3y + 11$

Exercises 9.3

1 (*a*) 1   (*c*) $z + 1$ (*e*) 2   (*g*) 16   (*i*) $42x$
2 (*a*) multiply by $a + 1$ (*c*) divide by 4
3 (*a*) 1   (*c*) $\frac{11}{7}$ (*e*) -4 (*g*) No number satisfies this equation.
    (*i*) No number satisfies this equation.

Exercises 9.4

1 (*a*) $2j$ (*c*) $j - 15$ (*e*) $7\frac{1}{2}j$ (*g*) $100 - j$
2 (*a*) $4d$ (*c*) $\frac{1}{12}d$
3 (*a*) c (*c*) $2c + 7$ (*e*) $200 - 3c$
4 (*a*) $A + 15$ (*c*) $2A + 3$ (*e*) 12
5 (*a*) $7c$ (*c*) $10c$ (*e*) 7 cents
6 (*a*) $2x$ (*b*) $2x - 10$ (*c*) \$22
7 Jones has 15 cows, Smith 30, and Brown 40.
8 $7\frac{1}{3}$ yards
9 21
10 24
11 19, 22, and 11 points, respectively
12 54 pieces
15 6 apples, 27 pears, 54 oranges
16 -5

Exercises 9.5

1 (b), (c), (f), (g), (i) are true, the others false.
2 $-2 = m$ and $1 = n$
3 (a), (b), (d), (e), (f) are true, the rest false.

Exercises 10.2

2 (a)

0.03	3.1	3.13
0.17	2.9	3.07
0.20	6.0	6.20

(c)

3.07	0.11	3.18
−1.11	2.03	0.92
1.96	2.14	4.10

3 (a) 400   (c) 14,878,000   (e) 42   (g) 42.04

4 (a) $4 \cdots, 3 \cdots, 12 \cdots, 12.587$

(b) $\dfrac{41}{10} \cdot \dfrac{307}{100} = \dfrac{12,587}{1,000}$

5 (a) $36\frac{12}{23}$, or about 36.52   (b) $4.596\frac{24}{31}$

6

	2.0	1.2	3.2
1.7	3.4	2.04	5.44
−2.3	−4.6	−2.76	−7.36
−0.6	−1.2	−0.72	−1.92

Exercises 10.3

1 (b) $\dfrac{21,875}{10,000} = \dfrac{5^4 \cdot 35}{5^4 \cdot 16} = \dfrac{35}{16}$

2 (a) .763   (b) 221

3 (a) 60 percent   (b) 85 percent

4 0.56%

5 $12\frac{1}{2}$ percent

6 (c) the commutative property of multiplication

8 (b) 10d   (c) 9 dimes, 27 nickels

9 (b) 0.666 $\cdots$; yes

10 (a) 0.142857 $\cdots$ (these 6 digits repeat forever)

```
(b) 7 · 2 = 1 | 4
    7 · 4 =       2 | 8
    7 · 8 =           5 | 6
    7 · 16 =              1 | 1 | 2
    7 · 32 =                      2 | 2 | 4
    7 · 64 =                          4 | 4 | 4 | 8
            1   4   2   8   5   7   1   4   2   8 ···
```

Exercises 10.4

5 (a) $\dfrac{2}{9}$ (b) $n = \dfrac{431}{999}$ (c) 1

Exercises 11.1

1 (*a*) 112   (*b*) 234
2 (*a*)   (*b*)

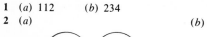

Exercises 11.2

1 (*a*) 23 + 12 = 40   (*b*) 32 + 24 = 111
2 (*a*) 143   (*c*) 121
4 (*a*) 23 − 11 = 12
5 (*a*) 221   (*c*) 221
6

31	31	112
10	34	44
41	120	211

8

	2	4	11
3	11	22	33
4	13	31	44
12	24	103	132

10 43
11 (*a*) 2   (*c*) 3   (*e*) 4
12 (*a*) 40   (*c*) 230   (*e*) 1,340
13 (*a*) 20   (*c*) 300   (*e*) 200   (*g*) 21
        (*i*) 141   (*k*) 201

Exercises 11.3

1 (*a*) final answer 113
2 (*a*) 2   (*c*) 26   (*e*) 116   (*g*) 2523
3 (*a*) 213   (*c*) 3132
4 (*a*) 12   (*c*) 300   (*e*) 400   (*g*) 13403
5 (*c*) five   12   sixteen   121
        six   20   seventeen   122
        seven   21   eighteen   200
        eight   22   nineteen   201
        nine   100   twenty   202
        ten   101   twenty-one   210
        eleven   102   twenty-two   211
        twelve   110   twenty-three 212
        thirteen   111   twenty-four   220
        fourteen   112   twenty-five   221
        fifteen   120   twenty-six   222

6 (*a*) 11   (*c*) 100   (*e*) 2120   (*g*) 21102   (*i*) 11201
7 (*a*) three   (*c*) twelve

8 (*g*) 100 always represents the square of the base
9 (*a*) 3   (*b*) 5   (*c*) 6
10 (*a*) 10   (*c*) 10   (*e*) 100   (*g*) 1000
11 (*a*) 1201   (*c*) 10103   (*e*) 716
12 (*a*) 13   (*c*) 286
13 (*a*) 12   (*c*) 27   (*e*) 120   (*g*) 2β

14 (*a*) $\frac{3}{7}$ (*c*) $\frac{5}{9}$ (*e*) $\frac{28}{42}$

15 Only (*e*) can be reduced, to $\frac{2}{3}$ (base five)

16 (*a*) $\frac{21}{22}$ (base five)   (*c*) $\frac{2}{14}$ (base five)

17 (*a*) 2.92   (*c*) 0.0176

Exercises 12.2

1 (*a*) 19.19   (*c*) 25.54 (these are approximate)
2 (*b*) The errors are, respectively, 0.0204, 0.0125, and 0.0002.
(*c*) In general the multiplier introduces errors of about 0.1 percent of the true answer, but sometimes the errors are smaller owing to favorable rounding off.

Exercises 12.3

2 (*a*) $\frac{1}{9}$ (*c*) $\frac{1}{125}$ (*e*) $\frac{1}{7}$ (*g*) $\frac{9}{4}$ (*i*) $\frac{25}{9}$ (*k*) $\frac{100}{3}$
4 (*a*) 10^3 (*c*) 2^3 (*e*) 5^{-3} (*g*) 10
5 (*a*) multiplication   (*b*) addition   (*c*) add
(*d*) subtracting   (*e*) $n - m$
6 (*a*) 10^{-3} (*c*) 10^{-2} (*e*) 10^{-9}

7 (*a*) (**1**) $\frac{47}{10^2}$ (**2**) $\frac{531}{10^2}$ (**3**) $\frac{230,007}{10^4}$ (**4**) $\frac{103}{10^1}$

(*b*) They are the same.

(*c*) (**4**) 5   (*d*) (**1**) $\frac{2,114}{1,000} \cdot \frac{30,027}{10,000}$ (**3**) $\frac{63,477,078}{10^7}$

Exercises 12.4

1 (*a*) 4.0×10^{11} (*c*) 2.997925×10^8 (*e*) 1×10^{-9}
2 (*a*) 2.111×10^5 (*c*) 1.311×10^7 (*e*) 3.14×10^0
(*g*) 8.95×10^{-5} (*i*) 1.21×10^{-1}
3 (*a*) 89.5   (*c*) 0.192   (*e*) 174,400   (*g*) 380

Exercises 12.5

1 Just over 479 feet
2 (*a*) 19.1884   (*c*) 45.2446   (*e*) 9,846.9

Exercises 13.2

1 (*a*) They are all equal, since their bases and heights are equal.
3 one-seventh
4

	A	B	C	D	E	F	G	H	I	J	K
Inside	0	0	0	0	0	0	0	0	0	2	2
Boundary	3	4	5	4	3	6	4	6	4	3	8
Area	$\frac{1}{2}$	1	$1\frac{1}{2}$	1	$\frac{1}{2}$	2	1	2	1	$2\frac{1}{2}$	5

In general, if *I* is the number of lattice points in the interior and *B* is the number on the boundary, the area is $I + \frac{1}{2}B - 1$.
5 (*a*) 25 square inches   (*c*) 13 square inches

Exercises 13.3

2 (*a*) Approximating π as 3.14159 yields a circumference of about 25,132.72 miles.
  (*b*) The clearance would be $100/2\pi$ feet. This is over 16 feet!
3 Circumference $= 2\pi$ feet, or about 6.2831858 feet. In 40,000 miles, tire makes about 33,633,000 revolutions

Exercises 14.2

1 34
3 51
5 48
8 123 miles
10 312 feet

Exercises 15.1

1 (*a*) 12   (*c*) 17   (*e*) 4
2 (*a*) 4   (*c*) 4   (*e*) −4

Exercises 15.2

1 (*b*) This is twice $\sqrt{2}$.
2 (*b*) $\sqrt{5} \cdot \sqrt{2} = \sqrt{10}$
4 (*a*) 9.2 feet   (*c*) 8.2
5 127.3 feet
6 (*a*) $\sqrt{9{,}879}$ feet or about 99.4 feet
  (*c*) $\sqrt{7{,}500}$ feet or about 86.6 feet
8 $\sqrt{136}$ inches, or about 11.66 inches

Exercises 15.4

1 No; these are all less than 1.5, while $\sqrt{3} > 1.7$.

2 (b) If $\frac{a}{b}$ is in the sequence, the next is $\frac{2a + 5b}{a + 2b}$.

   (c) $\sqrt{5}$

3 (a) These yield approximations to $\sqrt{10}$

   (b) These yield approximations to $\sqrt{17}$

4 (a) $\sqrt{101}$ (b) $\sqrt{197}$

Exercises 16.2

1 (a) 4   (c) 0   (e) $2\frac{1}{2}$ (g) $\frac{1}{6}$ (i) $\frac{5}{6}$

2 (a) 36   (c) 216   (e) 0   (g) 3,960

3 (a) 30   (c) 21,600

4 $32\frac{8}{11}$ degrees

5 15

6 (a) 21°   (c) 310°

8 684,000

9 (a) NE   (c) WSW

10 (a) 315°   (c) $337\frac{1}{2}°$ (e) $236\frac{1}{4}°$

11 (a) 35°   (c) 110°   (e) 200°   (g) 290°

12 240°

14 345°

16 120°

17 (a) 360°

Exercises 16.4

1 (a) 4 meters

2 (b) 115 kilometers

3 300 miles (rounded off to the nearest whole number)

6 $41\frac{1}{4}$ meters

8 $4

9 (c) 100 times the area of A (d) $x^2 : 1$

11 (c) multiply it by 49

12 (b) 27 units

Exercises 17.1

1 (c) yes

2 (f) no; see Fig. 17.1B

3 (a) 72°

4 (b) They are equilateral and all the same shape.

5 (a)

3	4	5	6	7	8
0	2	5	9	14	20

5 (b) 4,850

6 (b)

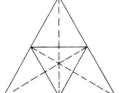

7 (*b*) square   (*c*) regular

8 (*a*) All except the kite   and the square.

9 (*b*) 72°
10 No; if a polygon has more than one line of symmetry, they all meet at a point.
11 (*b*) yes; 180°
12 (*b*) yes; 90°

Exercises 17.2

1 (*a*) (4,6,12) [(6,12,4) and (12,4,6) are also correct]

Exercises 17.3

1 regular tetrahedron
3 Use alternate vertices (as in Prob. 1) of the cube in Prob. 2.
4 (*a*) regular tetrahedron   (*c*) cube   (*e*) regular dodecahedron
7 regular octahedron

Exercises 18.1

1 (*a*) 60   (*b*) 2   (*c*) 60 ÷ 2 = 30   (*d*) 60
   (*e*) 5   (*f*) 60 ÷ 5 = 12
2 (*a*) 38   (*d*) 4 · 6 + 4 · 3 = 38

Exercises 18.2

1 (*b*) truncated tetrahedron   (*c*) hexagonal prism
3 (*d*) $R + V - E = 2 - 2H$, where H is the number of holes
4 (*c*) 100   (*e*) odd   (*g*) It is always even
5 (*a*) 120

Exercises 18.3

2 (*b*) At most 2
3 (*b*) no; all 20 vertices have degree 3
4 (*a*) no
   (*b*) yes; provided the start and end of the trip are at Wards Island and the Bronx.
5 (*a*) no   (*b*) Hamiltonian path

Exercises A.1

4 (*a*) They have gloves but not bats
  (*c*) They have bats but not gloves.
5 (*a*) 3   (*b*) 9
6 (*b*) 38
8 30

Exercises A.2

1 (*a*) {1,2,3,4,5}   (*c*) {$5\frac{1}{2}$}   (*d*) {2,3,5,7,11,13}
2 Only (*b*) and (*e*)
3 (*a*) {H,I}   (*c*) {D,e,b,F,g,A,C}   (*e*) {A,C,F,g}

Exercises A.3

1 (*a*) W and V   (*c*) S and V   (*d*) T, R, W, S, and V
2 (*a*) Ø   (*b*) {p}, {q}   (*c*) {p,q}
5 (*a*) 4   (*b*) 8
  (*d*) A set with *m* members has 2^m subsets. A set with 10 members has 2^{10} or 1.024 subsets.

Exercises A.4

1 (*a*) $A \cap C$ (*b*) $B \cup C$ (*c*) $A \cap B \cap C$
2 (*c*)   (*e*)

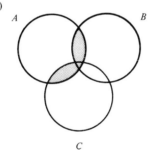

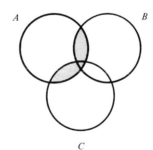

3 (*d*) {○ △ □ ● ▲ ■ •}
4 (*a*) 5   (*b*) 8   (*c*) 0   (*d*) 3

(*e*)	$A \cup B$	5	6	7	8
	$A \cap B$	3	2	1	0

5 (*a*) 116   (*c*) 37   (*e*) 33
6 (*a*) 27   (*b*) 10
7 (*a*) {*x* : *x* is a multiple of 6}   (*c*) {*x* : *x* is a multiple of 12}
  (*e*) {*x* : *x* is a multiple of 15}
  (*g*) {*x* : *x* is a multiple of *a*} ∩ {*x* : *x* is a multiple of *b*} = {*x* : *x* is a multiple of the least common multiple of *a* and *b*}
8 (*a*) {1,3,7,21}   (*c*) {1,3}
  (*e*) The answer each time is the factors of the highest common factor of the numbers whose factors were used to form the sets that were intersected.

INDEX

INDEX